CHILTON'S Guide to VCR Repair and Maintenance

CHILTON'S Guide to VCR Repair and Maintenance

Gene B. Williams
and
Tommy Kay

Chilton Book Company
Radnor, Pennsylvania

Published in Radnor, Pennsylvania 19089 by Chilton Book Company

Manufactured in the United States of America

Library of Congress Cataloging in Publication Data
Williams, Gene B.
Chilton's guide to VCR repair and maintenance.
Includes index.
1. Video tape recorders and recording—Maintenance and repair. I. Kay, Tommy. II. Title.
TK6655.V5W55 1985 621.388'332 84-45694
ISBN 0-8019-7606-5 (pbk.)

1986 Revised Printing: ISBN 0-8019-7785-1

9 0 4 3 2 1 0 9

ACKNOWLEDGMENTS

Special thanks to
Stewart Eisner and Ray Como
of The Federated Group
and to
Deke Barker,
who helped with this project.

Contents

CHILTON'S Guide to VCR Repair and Maintenance

Chapter 1
You Can Do It

As the owner of a video cassette recorder (VCR), you are no longer tied to the scheduling of network programmers. If a program you want to see is on at three in the morning, and you prefer to be getting some sleep, your VCR will take care of it for you. If both NBC and CBS are running blockbuster movies at eight o'clock Thursday night and you don't want to miss either of them, you merely push a few buttons, then sit back and enjoy one while the other is being taped. You can erase a tape at any time by recording over a program you don't want to keep permanently.

Beyond this, you can rent movies and watch the uncut versions, without commercials, anytime you wish. You can also buy movies and make them a permanent part of your movie library.

Another side benefit is that you are now in a position to get away from the expensive world of film for recording family vacations and events. Simply put a fresh cassette into the recorder, attach the camera, and make a permanent record of whatever you wish, complete with color and sound.

BUY ANOTHER?

Yet, all is not glamour and fun. Things can go wrong with the machine. The unfortunate truth is that your VCR is new only until you plug it in. The second that electricity flows in, the machine becomes used. A complete warranty lasts from 30 to 90 days, depending on the machine and the manufacturer. Even when the machine is under warranty, you might face certain charges.

Some years ago the cost of a simple transistor radio was high enough to make paying a technician for repairs worthwhile. Today repairs might total two or three times the cost of the radio. "It's not worth fixing," the technician will tell you. "You can buy a replacement for less than it would cost for me to repair it. Just throw it away and buy another."

A transistor radio is a relatively inexpensive device these days, so it's easy to understand why people throw them away when they break down rather than spend the money for repairs. It's not as easy to understand why someone would throw away a $500 VCR; however, the "Junk it and buy another" philosophy is common.

The rising cost of technical service is the prime contributor to the "Buy another" mentality. The prevailing attitude seems to be that repair of many pieces of electronic equipment isn't worthwhile. Efficiency experts claim that the cost of parts and technical service and the waiting time for repair often make it cheaper to buy a new unit if the original unit cost $600 or less. One person who uses video recorders in his work comments: "If the repair costs me $450, and I can get a brand-new machine for $600, why bother? For just $150 more I can get a new machine, complete with a new warranty."

Speeding us into this era of rapid obsolescence is the shortage of qualified service and repair technicians. Many people who are called "technicians" are in reality little more than "board changers." They can make a quick check of faulty electronic equipment and determine which circuit board is malfunctioning. The component that is causing the trouble might be a 5¢ resistor or a 25¢ capacitor, or perhaps even a $4 integrated circuit (IC) chip. Yet they make the "repair" by replacing the entire board, then charge you $100, $150, or $200, plus labor costs.

Swapping circuit boards is so quick and easy that most shops have set a minimum of one hour as a labor charge, regardless of how long the job really takes. Ten minutes of work, then, will cost you the full hour fee of $35 to $60. All this because there may have been a power surge in the line that blew out a 25¢ fuse.

In one instance, a VCR owner was having problems with her machine and brought it to a shop. She was told, " The heads have to be replaced. That will cost you about $375, including labor." The unit had cost her $430 originally, so she was ready to throw out the machine and get another. Fortunately, she decided to get a second estimate, and this time she managed to talk with a more honest technician. After a few minutes of examining the VCR, the technician said, "The heads are just dirty." Five minutes and $15 later, the VCR was functioning perfectly.

WHY THIS BOOK?

When you buy video equipment, you are given an owner's manual, which shows you how to use the machine. However, it rarely gives you the information you need to keep the VCR performing at its peak. For that, you are expected to rely on a service technician.

The VCR is a precision electronic-mechanical unit. Many different types of malfunctions may cause your unit to fail. These malfunctions often are a result of either component failure or operator misuse or neglect. Just as you maintain your automobile with periodic servicing, you must take care of your VCR. And as with your car, simple preventive maintenance will help avoid serious, costly problems.

Even a novice can be shown how to diagnose common machine malfunctions and make simple repairs, leaving only major repairs to high-priced service technicians. The more you learn about correcting minor problems, the more you will be able to save in time, money, and ruined recordings.

Without certain routine cleaning and adjustments, your VCR is quite likely to begin producing poor pictures or even go completely dead in less than two years. This book will help you prevent this from happening, even if you have no prior technical or mechanical background. You don't have to be at the mercy of the repair shop, and you certainly don't have to toss $500 or more of equipment in the trash. If you have enough "skill" to replace a burned-out light bulb or use a screwdriver, you have sufficient background to use this book.

In addition, you'll learn how to recognize when the malfunction is best left to a professional technician. When and if this time comes, you'll have some idea of what has gone wrong. Your knowledge will help reduce both the cost of repair and the chances of being ripped off by a dishonest or less-than-competent technician.

Most VCRs have a sticker that says something like, "WARNING—SHOCK HAZARD—DO NOT OPEN! No user-serviceable parts inside. Opening this cabinet will void all manufacturer's warranties" (see Figure 1-1). Even after the warranty has expired, this tag deters the owner from attempting even the simplest repairs.

With proper precautions, there is little danger under the cover (Figure 1-2). Unlike a television set, a VCR operates at low voltage. The most dangerous spot is where the 120-volt supply comes into the

FIG. 1–1 The warning sticker: Let it remind you to be careful.

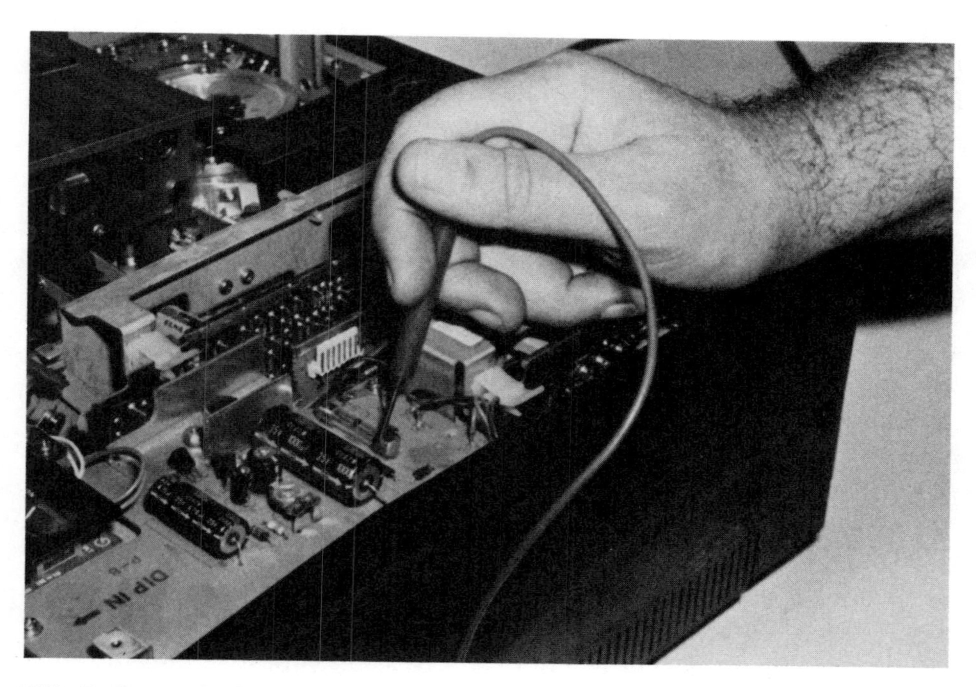

FIG. 1–2 Inside the VCR.

unit. By carefully following the instructions in this book, you will never face danger from this spot. (See Chapter 2 for more details.)

TOOLS YOU'LL NEED

You probably already have most of the necessary tools. Even if you must purchase them, the cost is usually low. Most are available at any electronics supply house or drugstore. The tools needed are the same for any make or model, whether front loader or top loader. The tools are described in the order of importance, beginning with those you'll definitely need and progressing to the more complex, optional ones. Table 1-1 lists approximate costs; "RS" denotes that the part is available at Radio Shack at a competitive price.

TABLE 1-1. Required Tools

Part	*Part Number*	*Cost ($)*
Screwdrivers (3 blade)	Any	3.00-6.00 each
Screwdrivers (2 Phillips)	Any	3.00-6.00 each
Needlenose pliers	Any	3.50
Regular pliers	Any	3.00
Hex wrench set	Any	2.00
Nut driver set	Any	5.00
Denatured alchohol	Technical grade	1.00-3.00/quart
Cotton swabs	Any	1.50
Head-cleaning pads	Any	2.00
Head-cleaning fluid	RS 44-1170	2.99
Head demagnetizer	Any	Varies
Contact spray cleaner	RS 64-2322	1.99
Beeswax	Any	Varies
Light machine oil	RS 64-2301	1.49
Tape tension gauge	Any	Varies
Torque spring	Any	Varies
Small magnifying glass	Any	1.00-5.00
Multimeter (VOM)	RS 22-201	19.95
Soldering iron (20-25 watt)	Any	Varies
Solder (resin-core only)	RS 64-001	.89
Desoldering tool	RS 64-2086	1.99
Heat sinks	Any	Varies
Circuit-board holder	RS 64-1801	6.95
1/32-inch drill bit	Any	Varies
Spare cables	Any	7.00
Spare fuses	Any	.25+ each
Spare belts	Any	1.00+ each

Figure 1–3 shows tools that will help you in your repair and maintenance work. You'll need screwdrivers to remove the cabinet and disassemble certain parts inside. You should have at least three blade-type screwdrivers of different sizes and two Phillips-type screwdrivers, also of different sizes. The cost isn't important, but the screwdrivers should be sturdy enough to turn the screws without bending, breaking, or flaking. (Less expensive screwdrivers are often coated with a substance that flakes off. Don't use them!) The blade of the screwdriver should fit nicely into the slot of the screw. Trying to use an inappropriate-size screwdriver can damage the machine.

A needlenose pliers is helpful for a variety of tasks, such as reaching into small places and retrieving parts you have dropped. Regular pliers, in a small size, are great for holding things your fingers can't. Both types should have insulated handles to protect you and the machine. As with the screwdrivers, be sure to get tools that won't flake.

A set of hex wrenches may be required, depending on the machine you own. A complete set is better, and less expensive in the long run, than individual wrenches. Again, beware of cheaply made tools that may flake or bend.

A set of nut drivers is not essential but comes in so handy that we chose to list it at the beginning. Nut drivers are like socket wrenches attached to a screwdriver handle. They are used to remove or install the various nuts that hold things together.

Denatured alcohol and cotton swabs for cleaning the tape guides are available at your local drugstore. DO NOT use isopropyl rubbing alcohol; it may contain oils that can harm the video equipment. Ask for *technical-grade* isopropyl alcohol. The cotton swabs can be any type, as long as the cotton is wound tightly so that it won't leave threads inside the machine.

A bottle of alcohol will cost about a dollar and will last for hundreds of cleanings. The cotton swabs are about a penny each. The total cost of cleaning the tape guides in this way is just a few cents. DO NOT use cotton swabs to clean the heads. They should be used only for cleaning the tape guides and other mechanical parts. (For more information on cleaning the heads and other parts inside the VCR, see Chapter 7).

Commercial head-cleaning kits provide convenient cleaning and cost from $10 to $20. A head-cleaning cassette can save you time and bother, but you'll pay as much as a dollar per cleaning. Many technicians don't recommend using cleaning tapes because they feel they do only a partial job of head cleaning, and are inefficient in tape-guide

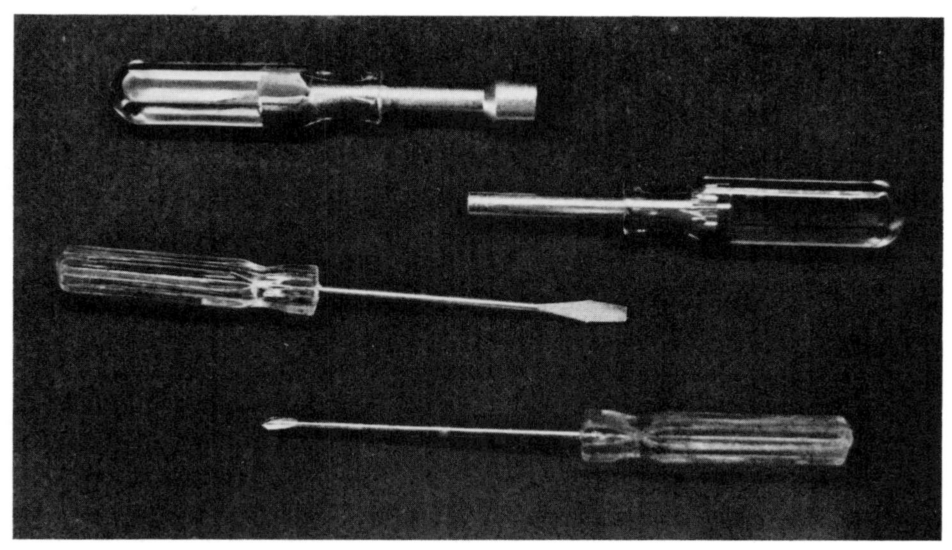

A

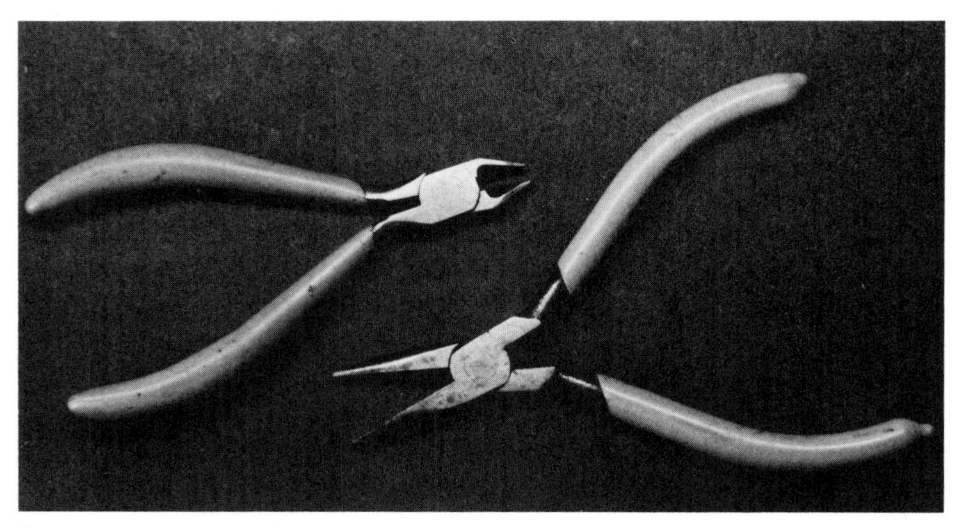

B

FIG. 1–3 (*A*) Screwdrivers and nut-driver set (*B*) Needlenose pliers.

cleaning. Further, some specialists believe that the cleaning tapes can even leave a harmful residue on the tape guides and the head drum, which can swiftly lead to a sharp deterioration in the performance of your VCR. It is best to clean the heads by hand (see Chapter 7).

Despite regular cleaning, the heads may become magnetized, just as do the heads of any recording or playback equipment that uses magnetic tape. To get rid of this magnetic charge, you'll need a head demagnetizer (see Figure 1-4). DO NOT use a standard audio head demagnetizer: because of its strong magnetic field, it could shatter the heads. (See Chapter 7 for more information on head demagnetizing.)

If your machine has plug-in circuit boards, you'll want some kind of contact cleaner, such as denatured alcohol and swabs. You can also use a freon-based cleaner or any commercial electronic cleaner. In a pinch you can even use a soft pencil eraser. DO NOT use a cleaner that contains a lubricant, because this will leave a residue on the contacts.

Occasionally the belts have to be "dressed," or restored, with beeswax. This can be purchased for very little at hardware stores and most drugstores. (*Note:* Dressing the belts is only a temporary cure for slipping. If you have to use beeswax to dress the belts, it's probably time to replace them.)

The first time you open your machine for routine cleaning and maintenance, check the number and size of all drive belts. Keeping a supply of spare belts can save a lot of time and trouble, especially if your machine malfunctions during the evening or on a holiday or weekend, when replacement parts may not be available. Keep the belts in their packages and out of direct sunlight.

A small can of light machine oil will be needed for lubricating some of the moving parts. You can get it from many sources, including small appliance stores. DO NOT use ordinary automotive motor oil.

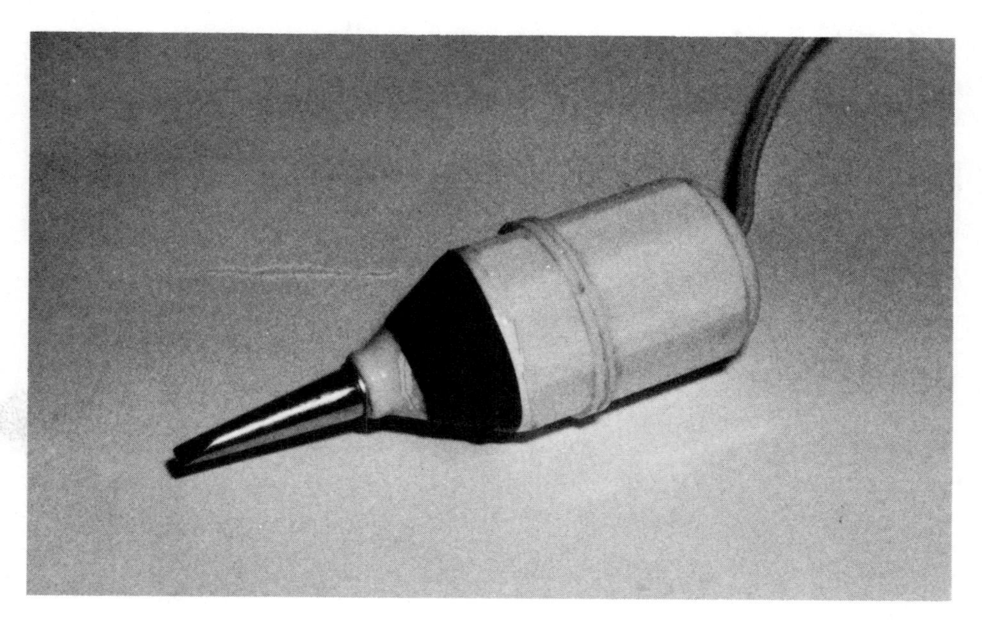

FIG. 1–4 Head demagnetizer.

To make internal adjustments, you'll need two special tools: a tape tension gauge to ensure that the tape is being held and fed at the proper tension and a torque spring to make various drive-mechanism adjustments. Both are inexpensive and available at any electronics supplier. You may need special versions of these tools for your particular machine, however.

A multimeter, or volt-ohm-milliammeter, called a VOM (see Figure 1-5), is used to check voltage, resistance, and current (under certain conditions). Although you can spend several hundred dollars for a VOM, one costing $20 or less will be fine. A VOM can also be used to check wall outlets or even the wiring to your stereo speakers.

For any technical work inside the VCR, you'll need two tools, shown in Figure 1-6. One is a soldering iron (10–25 watt). If you're going to be doing any soldering, learn correct procedures. Soldering is not as easy as it sounds, and you can cause considerable damage if you're careless or inexperienced. The other is a desoldering tool, sometimes called a "solder sucker" (which describes its function well). These tools come in several different types, from a simple squeeze bulb to a fancy syringe. Virtually all electronic parts in a VCR are soldered into place. If you have to change one, first you must melt the old solder joint. Then you use the desoldering tool to remove the melted solder. This makes component removal much easier. Some components, such as IC chips, cannot be removed without this tool.

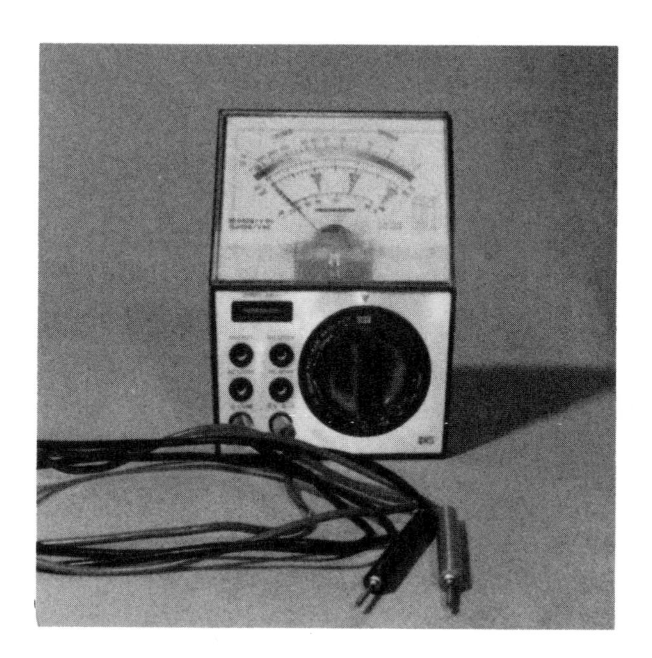

FIG. 1–5 A multimeter (VOM) is essential for troubleshooting.

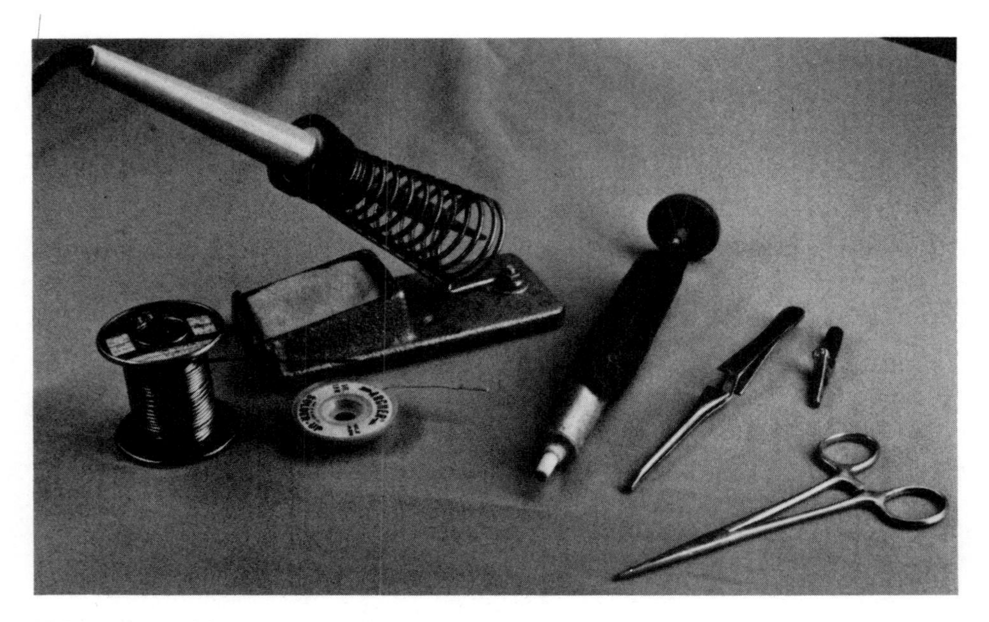

FIG. 1–6 Soldering iron, resin-core solder, desoldering tool, and heat sinks.

For safer soldering, a set of heat sinks (see Figure 1–6) is helpful. These are small metal clamps that draw the excess heat away from the circuitry and thus protect your machine. Heat sinks also help protect new components while you solder them in place. A component such as a transistor or IC chip is sensitive to heat. Pliers, such as a needlenose, can also be used, but the heat sinks are much better because they don't require that you hold them in place.

Since VCR components are quite small, you can easily burn your fingers as you solder or desolder them. When carrying out some repair jobs, you will probably find yourself wishing that you had an extra hand or two. A small C-clamp or vise or a special circuit-board holder will serve as another hand (see Figure 1–7). The saving in frustration alone is worth the eight- or nine-dollar cost. Be sure to pad the jaws if they're not already padded. (A folded paper towel will do.)

A $\frac{1}{32}$-inch drill bit can be used to clean out clogged circuit-board holes. You may even be able to find a drill that works as a soldering-iron attachment at one of the larger electronics supply stores. One of the small hobby drills, such as the Dremel MotoTool, is better for this job than a standard hand drill, but a hand drill will work if you're careful. Most of the time you won't actually be drilling anyway—you'll be using the drill bit to clear a circuit-board hole that is clogged with solder.

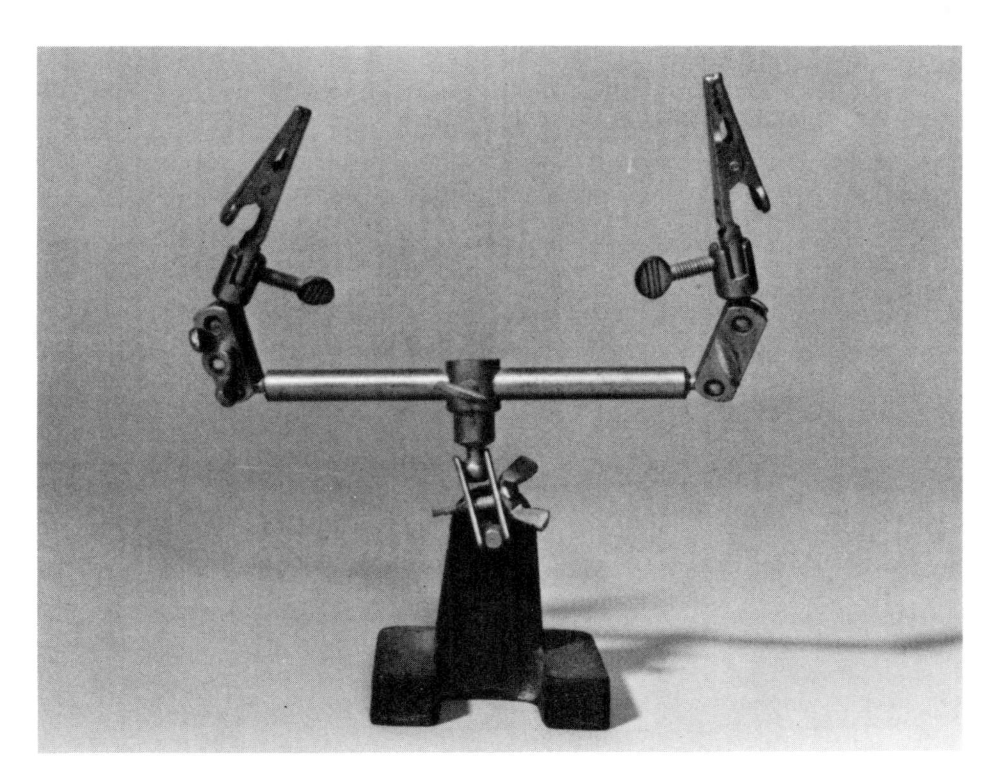

FIG. 1–7 A "Third Hand" is a worthwhile investment.

If possible, keep some spare cables and connectors on hand for external connections. The cost is small compared to the frustration of having a cable break just as you're about to make an important recording.

SUMMARY

With the required tools, you can take care of many VCR repairs and all regular maintenance. Taking care of a VCR is more a matter of common sense than anything else. If you can read and follow basic instructions, you can increase the life of your VCR and save considerable time and money in repair. You will rarely have to throw away a malfunctioning machine.

The first step is to learn how to maintain the equipment properly. If you spend just a little bit of time in maintenance, the number and seriousness of malfunctions will be greatly reduced. This maintenance covers the machine and the tapes.

Before beginning any of the procedures, be sure to read the instructions thoroughly and follow each step exactly as described. Go through the steps mentally before you actually perform them. This is especially important if you've never attempted to maintain sophisticated equipment before. There's no reason to fear the equipment or the repair of it. By going slowly and thoughtfully, you will reduce the chances of accident and increase your understanding of the machine.

Chapter 2

Safety and Preparation

If you're going to be working around electricity and complicated equipment, the first thing to learn is how to proceed safely. Although most parts of the VCR present minimal risk to you, the voltage and current in certain spots can be dangerous—even fatal. Basic precautions and common sense will see to it that neither your safety nor the VCR's is jeopardized.

YOUR SAFETY

The first and most important consideration is your safety. If you damage the VCR during repair, a professional can come to the rescue, and if he or she cannot salvage the machine, you can replace it. However, *you* cannot be replaced.

Every year a number of people are accidentally electrocuted. Almost without fail, the cause was simple carelessness—the notion that safety is for "the other guy."

There is no such thing as being *too* cautious.

Most electronic equipment that receives current from a wall socket has a sticker warning you to stay out. The warning is often accompanied by "No user-serviceable parts inside." The purpose of these stickers is to discourage people from poking around where they might come into contact with a dangerous amount of voltage.

Warnings carried on most television sets say that even when the unit is unplugged from the wall socket, probing inside the chassis may expose you to a potentially lethal shock. The operation of a video tube in a television set requires extremely high voltage. Even with the set turned off and the cord unplugged from the wall outlet, the electronic devices inside the set retain a hefty charge. Capacitors in particular can store large and dangerous voltages. The stored charge can last for hours—even days—after the set has been turned off and unplugged. Sticking a finger in the wrong place inside a television set can be instantly fatal.

A VCR does not have a picture tube and therefore does not need to generate such high voltages. The 120 volts AC from the wall socket operates the motors and servomechanisms for the various functions of the VCR. This voltage presents the only threat to your safety (Figure 2–1). With proper precautions, working on a VCR is no more dangerous than working on an electric lamp.

Still, the dangers of electricity cannot be taken lightly. To find out what events take place in an electric shock, the United States Navy performed tests with the standard 60-cps (cycle per second) alternating current (AC) that flows through the power lines, out the wall outlet, and into the equipment (such as your VCR). These tests have shown that it takes just a tiny amount of current to kill. A current value of just one milliamp (.001 amp) can be felt. A current of 10 milliamps (.01 amp) causes the muscles to become paralyzed, making it impossible for the person to let go of the source of the shock. A current of 100 milliamps (.1 amp) is usually fatal if allowed to continue for more than one second.

FIG. 2–1 Be especially careful when you work near where the 120 volts AC comes into the VCR.

The amount of incoming current is limited only by the wiring and the circuit breakers or fuses. Normally, the current is a steady 15 or 20 amps. For a short time, until the wires melt or the circuit breaker or fuse blows, the current is almost limitless. If you are careless, your body could absorb several thousand times the lethal amount of electrical power.

Safety Rules

1. ***Remove all jewelry before starting.***
2. ***Use the one-hand rule.***
3. ***Insulate yourself from the equipment.***
4. ***Insulate yourself and the equipment from the surroundings.***
5. ***Work slowly and carefully.***

Technicians are taught two precautions on their first day in training school. The first is to remove all jewelry, watches, belt buckles, and other metallic objects from the hands, wrists, and neck before attempting to do any electrical service work. Not only can a dangling bracelet or necklace conduct electricity and give you a serious shock, but it can also act as a conductor between a hot wire and the chassis or another surface and cause a short, which can seriously damage the unit.

The second precaution is to follow the one-hand rule. Keep one hand in your pocket at all times during investigation or probing of an electronic device that is energized—that is, has current flowing in it. This suggestion at first may sound strange, if not silly. Yet electricity, to shock, must flow from one point to another. Current will flow from a finger than touches an energized wire to another spot. If one hand touches the metal chassis of the VCR or another conductive surface and a finger of the other hand touches a high-voltage line, current will instantly pass from one hand to the other, and through your body.

If you keep one hand in your pocket, current cannot flow through your body from hand to hand after you carelessly touch a conductive surface. By keeping the free hand in a pocket, you won't be as inclined to reach out and touch something you shouldn't.

The third and fourth rules are to insulate yourself and the equipment. Wearing rubber-soled shoes and working on a unit that is placed on an insulated surface (e.g., a dry, nonmetallic table) will prevent a dangerous current from flowing through your body.

VCR SAFETY

The power supply changes the 120 volts AC to the values of direct current (DC) needed to operate the circuitry. The voltages used for amplification, tuning, and basic circuit functioning of the VCR are generally very low, with 5 or 12 volts DC the most common. Even with the VCR plugged in and the power on, these circuits present no danger to you. However, they can be dangerous for the VCR. If your screwdriver or another metal object touches two wires, you could cause a short circuit. The circuits are designed to handle a certain amount of current. With a short circuit, a component that is meant to carry just 1 or 2 volts (or less) may suddenly be hit with 12 volts and a higher amperage.

A friend was probing the power-supply outputs. His hand slipped and the probe touched the wrong spot. Before he even realized that his hand had slipped, an entire circuit board was ruined.

Always turn off the power to the unit while attaching test leads, especially if you've had little prior experience with a VOM. Apply power to take the reading, then disconnect the power before removing the leads. This will help prevent accidental shorts.

In addition to electronic damage, there is the possibility of physical damage. Some VCR parts are delicate, so keep your fingers and anything else that might be contaminated with dirt or grease out of the machine. The oils from your skin can ruin certain parts as easily as a hammer.

The rotating record/playback heads, for example, are extremely sensitive, especially when they're moving. Your fingers should NEVER touch the heads. (And there is no reason for them to, since head replacement must be left to a professional technician, who can align a new head properly.) The only things that should *ever* come in contact with the heads are cleaning fluid and pads.

Whenever you use a demagnetizing tool, make sure that it never touches the parts being demagnetized. For example, the head demagnetizer must NEVER come in direct contact with any part of the unit, especially not the heads. The tool's magnetic probe vibrates when energized and will shatter the heads if it touches them. The magnetic energy may be strong enough to create oscillations that can damage the heads. The demagnetizer should never be held too close to the head and tape guides. Chapter 7 gives you more details on how to use a head demagnetizer.

Haste is your biggest enemy. *Go slowly.* Plan what you are going to do before you do it, and make sketches to help.

VCRs are designed to be operated in a horizontal position. If the power is on with the unit tilted on its side, some of the motors may be forced out of line by their mere weight, causing drive belts, springs, levers, or servomechanisms to slip out of place. The machine could jam or be damaged.

Keep movement of the VCR unit to a minimum, during both repair and routine operation. This precaution is clearly more important with tabletop models than with portables, but all VCRs should be treated with care, especially while power is flowing through them.

A final consideration concerns positioning the machine for adequate air circulation to prevent overheating. Most units have air vents at various locations on the cabinet. These vents allow the release of heat generated by the components. Often an unwary user will set the machine on a table or stand too close to another object or in a position that blocks the air vents. The result is overheating. And the result of overheating is almost always a damaged unit.

One user was so concerned about dust getting into his machine that he covered all the air vents with masking tape. About a week later, he stood, burned-out unit in hand, wondering why the manufacturer would build such an unreliable product.

REMOVING THE COVER

Few pamphlets, instruction books, or manuals tell you how to open your VCR. They may tell you exactly what to do once you're inside, but how can you do anything unless you can get inside?

Opening the cabinet is relatively simple. Inside the machine, however, there are various "hot spots" where you could accidentally touch a live current. Therefore, the power MUST be off and the plug disconnected from the wall socket *before* you attempt to remove the front or back cover. These two precautions will help protect you from potential shock and your VCR from possible damage.

Frequently the cover is in more than one piece. Usually you can remove the top cover without danger of accidentally loosening screws that hold other parts in place. The bottom cover might present more of a problem, because it contains screws that secure other components, such as the power supply. So check carefully before you remove any screws.

Warning

Removing the cover may void the warranty. Carefully read the warranty information that comes with your machine before you remove the cover.

To remove the top cover of your top-loading VCR, first unscrew all fasteners that hold the cover to the main unit. Then carefully try to lift the cover (Figure 2–2). If it does not move, run your fingers around the edges to see where the cover is still attached. You may find that you must remove a knob before the cover will lift off over the shaft. Or a screw at the side may be securing some kind of brace to the cover. **NEVER** force the cover. A top-loading VCR allows at least some access to the interior. A front-loading machine is basically sealed; however, once you have the cabinet off (Fig. 2–3), the differences between a front loader and top loader are minor.

Remove the top cover as soon as possible after the warranty has expired. By looking inside as it operates, you can gain a better understanding of how your unit does what it does. Look at the connectors

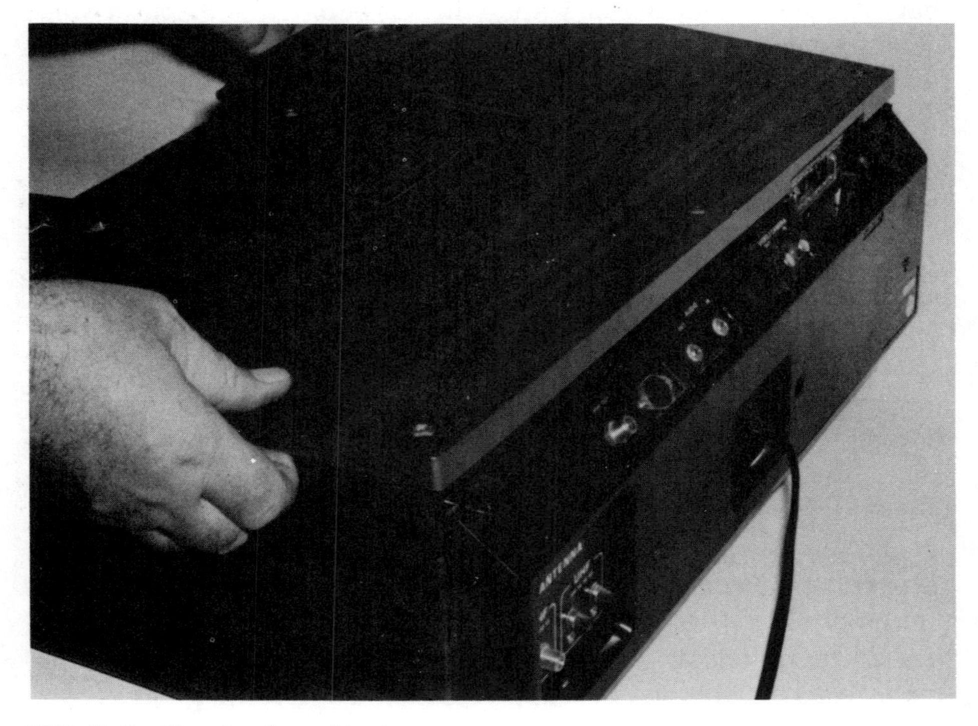

FIG. 2–2 Opening the cabinet.

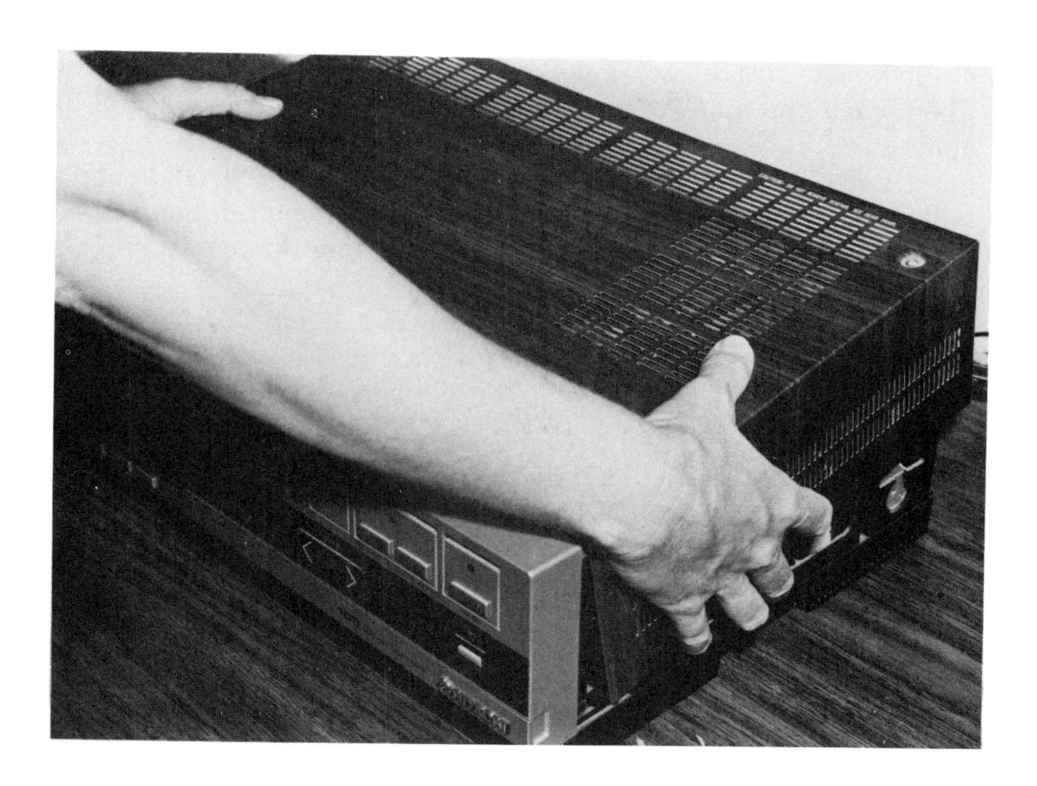

FIG. 2–3 *Top*, removing the top cover of a front-loading VCR; *bottom*, inside the cabinet of a front loader.

on both sides of the back panel as you perform the initial hookups. Then when you apply power and insert your first cassette, you'll be able to see the automatic threading machinery in action.

Often when the first-timer opens the VCR, screws fly out haphazardly. Once the cover is removed, the owner is amazed that components have fallen out. And when it comes time to reassemble . . . "Now where does this go?"

The key to successful repairs is care. Plan every step before you take it. Whenever applicable, make sketches and notes. Keep the parts you remove organized, in a muffin tin, for example.

PREPARING TO WORK

Preparation is a crucial part of safety. Understanding your machine is the best way to prepare for maintenance or repair work. If you understand how your unit operates, you'll be more likely to produce safe, successful, and timely results.

With the machine upside down and the bottom cover removed, study the positioning and alignment of the various motors and belts used to drive the mechanisms of the VCR. Referring to your owner's manual or service manual, mark down the position and size of each belt, drive wheel, pulley, and fuse (Figure 2–4) and note how all gears and pulleys engage so that you will be able to obtain and install replacements when necessary. If you open the VCR for the first time only after you have encountered trouble, you may find it difficult to determine, for example, the correct tracking position for a replacement belt. While the unit is open, try to determine the proper method of removing and replacing the mechanical parts, belts, and fuses.

During this inspection, be careful not to apply pressure at any point on the mechanical parts of the VCR. The alignment of most of the levers, tape guides, and servomechanisms as well as the tracking wheel is fairly critical. A piece that is bent even slightly out of shape could cause the unit to malfunction.

Next, check the tension of the various belts and springs (Figure 2–5). After many hours of operation, these belts, like the fan belts of an automobile, become worn and stretched. When a belt becomes worn, it no longer maintains the proper tension to operate the mechanism for which it is designed. Slightly stretched or worn belts may be temporarily helped by an application of resin or beeswax to the inner surface (Figure 2–6) to prevent slipping. Dressing the belts in this way may help out in a pinch—and allow you to finish watching a movie or to make an important recording.

FIG. 2–4 Note the position, size, and alignment of all belts, gears, pulleys, and fuses.

FIG. 2–5 Checking the tension and condition of a belt.

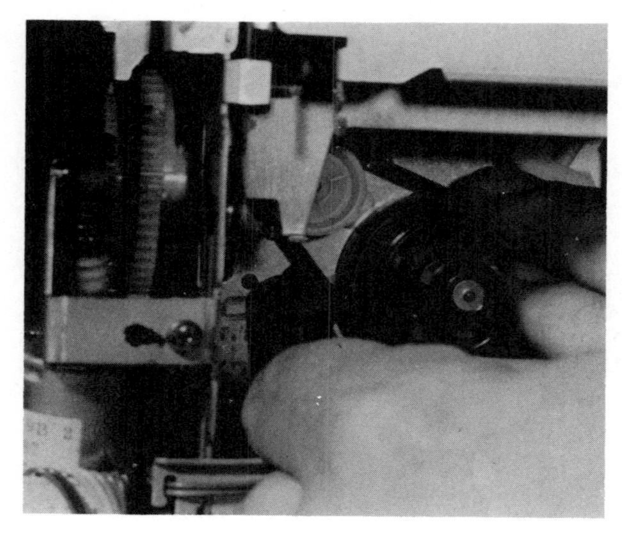

FIG. 2–6 Applying resin or beeswax to a belt.

Note the location of all fuses and breakers when you make your initial study of the VCR. Fuses are often concealed so that they are virtually inaccessible. However, you don't need to contribute to the service technician's next trip to the Bahamas: *You* can locate and replace a fuse.

Two commonsense alerts are worth restating here: NEVER replace any fuse with a wire or with a fuse with higher current breakdown value. If the fuse is rated at 1 amp, replace it *only* with another 1-amp fuse. Failing to follow this precaution could lead to more serious damage to the machine and even create a hazard to you. (Electrocutions and house fires result when people don't bother to use common sense in replacing a fuse.)

Always turn off the power and unplug the machine before changing a fuse. Chances are good that a 120-volt current is going to the fuse junction block. Sticking a finger or screwdriver there while the machine is plugged in is like sticking it into a live wall outlet.

If the fuse is held tightly or is not easily accessible, a fuse puller (see Figure 2–7), which costs approximately a dollar, can help.

As you study the back of the VCR, pay close attention to how the various external connectors are mounted. Every time a cable is attached to one of these connectors, twisting or pushing force is applied. Eventually, this movement may cause one of the inner wires to break or become shorted. If a VCR malfunctions and the problem, through testing, is tracked to one of these connectors, look inside and check for a broken wire or a wire that has been twisted into a shorting position (see Figure 2–8). You may be able to make the repair simply by untwisting or resoldering the wire, rather than having to replace the entire connector. Also make sure that the nuts mounting these connecting sockets are firmly screwed down. They may have become

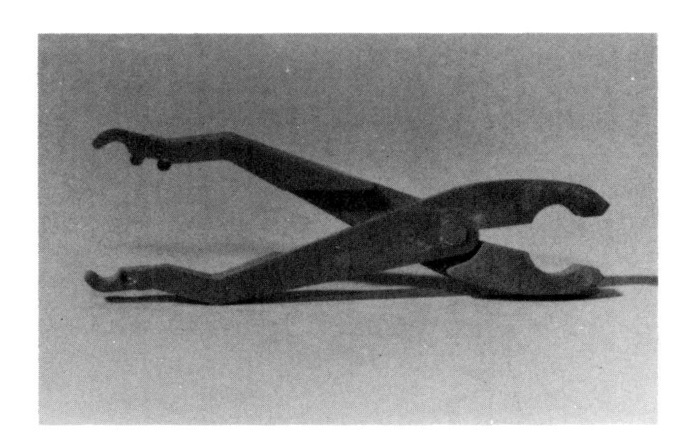

FIG. 2–7 A fuse puller.

FIG. 2–8 Check the connectors to see if any wires are broken, loose, or twisted.

loosened, either through overuse of the connector or during transport of the VCR.

After you have thoroughly studied the interior of the unit, replace the bottom, side, and back covers and place the machine in its normal operating position. Plug in the unit, turn on the power, and study the tape transport and tape guide system. Load a cassette and watch the self-threading mechanism. Several of the levers on this mechanism are made of a fragile plastic material or thin metal. These are all too easily bent or broken. See how they are *supposed* to operate so that in case one of them does break or bend, you'll be able to pinpoint the difficulty and install the replacement part.

KNOW YOUR LIMITATIONS

Although you can handle more than 75 percent of all repairs and maintenance, some repairs are best left to professionals.

You will most likely be able to handle such basic malfunctions as blown fuses; worn drive belts, pulleys, or pinch rollers; malfunctioning connectors or broken interconnecting cables; and dirty, clogged, or bent tape guides. (These are the areas of the machine that should be studied most carefully when they are operating. This way you'll know what to look for when a malfunction occurs.)

For tape alignment problems, however, and for repair of the complex electronic circuits, you will probably have to call a professional serviceperson. Alignment is one of the most common things for which you'll require a professional technician. Fortunately, you don't need to worry about head alignment very often. Alignment is done during manufacture and isn't needed again until about 2,000 operating hours have passed. That's like watching or recording 1,000 two-hour movies. If you watch a double feature every night of the week, it will take almost three years before you reach 2,000 hours.

Professionals usually use a special (and expensive) alignment tape, along with some associated (and expensive) test equipment, to check and then touch up the VCR's playback/recording quality. This equipment is *not* used for troubleshooting, since the alignment tape can be damaged all too easily by a malfunctioning machine. Alignment tapes are usually available only to authorized dealers.

Since alignment is needed so seldom, and since it requires expensive equipment that's not readily available, leave the job to a professional.

FIG. 2–9 The circuit boards can be damaged by heat.

You should also leave to the professional anything that is too complicated for your experience. For example, some of the special options, such as digital timers, require special equipment to test or repair.

Unless you know how to handle a soldering iron correctly (Figure 2–9), leave any complicated soldering to a professional. The boards are sensitive to heat and can be damaged easily. The same is true of certain components, such as the IC modules.

COMPONENT REPLACEMENT

Sometimes you'll be able to track the malfunction of your VCR to a single component. Perhaps a five-cent resistor has burned out, or perhaps a thirty-cent transistor is faulty. You don't necessarily have to throw away an entire circuit board.

Repair on a component level requires that you have experience with soldering. If you don't, either learn how to solder and practice *before* you attempt any repairs or, better, leave all soldering jobs to a professional.

Desoldering and soldering of most two-lead components (resistors and capacitors) are fairly simple. Keep in mind that the circuit can be damaged—permanently—by too much heat. Make sure that the soldering iron is never in contact with the board for more than five seconds.

Working with components that have more leads can be tricky. Components that use more leads tend to be more complex and thus more prone to heat damage. It isn't difficult to destroy a transistor with too much heat. At the same time, too little heat will make a weak solder joint. The circuit will operate sporadically at best, and you may damage other components.

IC modules are sensitive to static as well as to heat. Handle them with great care, and avoid touching the leads with your fingers or with uninsulated tools. IC extracting and inserting tools make the job of replacement easier. If you must solder around the IC modules, use one of the special soldering tools designed for electronic circuits. These have grounded tips that prevent static from building up on the tip and damaging sensitive components.

Component replacement requires an exact match. If you're not sure how to "read" the value of a component, get one of the many books available on basic electronics.

Some VCR components have polarity. Installing such a component incorrectly can cause severe damage. In some cases, it can also cause a dangerous explosion. Electrolytic capacitors and some IC chips are well known for this.

Diodes and electrolytic capacitors are examples of polarized two-lead components. The transistor is probably the best example of a three- or four-lead component with polarity. IC modules can have any number of leads coming from the package, and the placement of each in the circuit is important. Pay close attention to the polarity of the component you remove. Even if the circuit board indicates the polarity, take notes and make sketches as you go along.

THE MOST IMPORTANT FACTOR

Cleanliness is essential to long life and top performance of a VCR. Even with proper maintenance, your VCR won't work perfectly forever, but by keeping the unit clean inside and out you can minimize the number of malfunctions that will occur. The majority of VCR malfunctions are the result of a dirty machine. Although the amount of dust that can get inside is reduced with a front-loading model (one of the major reasons for growing popularity), some dust will get into any model.

Keep ashes, food particles, and liquids away from the machine. Dust and dirt or spilled coffee can do tremendous damage. Never operate the unit when your hands are even slightly dirty. Also keep the area around the VCR as clean as possible. A dust cover is an inexpensive investment, and a valuable one. If a dust cover didn't come with the VCR when you bought it, buy or make one.

Cleanliness doesn't stop at the machine. Dirty tapes can also cause considerable damage. Dust, ground-in dirt, or grease on a tape causes dropouts in the picture and eventually can damage an entire tape cassette (see Chapter 6).

Even if you are particularly careful in how you use and store your video recordings, you still stand the risk of a dirty tape damaging your machine. A tape can pick up dust particles and grease from one machine and transfer them to another. This is an especially important concern if you plan to rent or borrow tapes. If they have been played on grimy or poorly maintained machines, dust and dirt will end up in your machine.

MAINTENANCE

The best way to handle problems is to avoid them. Regular maintenance will take care of this and will greatly reduce the cost, both in time and money, of keeping your VCR working properly. (See Chapter 7 for a more complete discussion of maintenance.)

To clean your VCR successfully, you must know which areas to clean. While the VCR is apart, study every point where the exposed tape touches a guide or part of the transport mechanism. These are the areas where anything bent slightly out of shape or contaminated with even minute particles of dust or grease can cause infinite harm. These are the areas, along with the record/playback heads, that should be cleaned routinely with either head cleaner or denatured alcohol. Do this after every 20 to 25 hours of operation.

It's a good idea to clean these mechanisms before making your first recording, since even while the VCR was in storage dust or moisture could have built up in these critical areas. (However, if you open the cover, you may void the warranty.)

SUMMARY

Use caution whenever working around electricity. Even currents that can't hurt you can severely damage electronic circuits.

Wear rubber-soled shoes, place the machine on an insulated tabletop or surface, avoid water or any type of moisture, and use an insulated probe. Always follow the one-hand rule to prevent your body from becoming a target for electrons.

Work slowly and carefully. It's better to take an additional half-hour to complete a job than to spend an entire day trying to make up for mistakes made in haste.

Take the time to learn how your unit operates and you will be less likely to run into trouble later. As soon as possible, or after the warranty expires, remove the top cover and watch the VCR in operation.

Learn to recognize your own limitations. Although you can easily handle the majority of problems that come up, sometimes it will be best—and less expensive—to pay a professional.

Chapter 3

VHS or Beta? New or Used?

VHS OR BETA?

Two basic design formats have been battling for consumer acceptance since the invention of the VCR. (*Format* refers to the VCR's general plan or arrangement.) Ask around and you'll find people, including highly qualified technicians, who will speak vehemently for and against both systems.

The Sony Corporation was the first to mass-produce VCRs. Engineers at Sony invented the Beta format. Soon after, the RCA Victor Corporation began to distribute the VHS format. Today, all VCRs sold in the United States are made in one of these formats.

The two formats are not compatible. You can't use a Beta cassette in a VHS machine. Each has different characteristics. For one thing, the VHS cassette is slightly larger than Beta, so it is impossible to load a cassette of one format into a machine of the other.

The main difference between Beta and VHS lies in the number and arrangement of tape-track guides. These track guides keep the tape stretched tightly in the correct position as the tape moves across the rotating drum, which carries the record and playback heads. The guides, along with the VCR's motors, drive belts, and other mechanisms, maintain the torque necessary to keep the tape moving around the head drum at the proper tension.

The VHS system has nearly twice as many tape guides as the Beta format. This makes the automatic threading of the VHS machines slightly more intricate.

So, which format is better?

In the past few years, the majority of units purchased have been in the VHS format. For this reason, many stores that rent prerecorded movies and other cassette materials to the public handle only VHS. For the same reason, there are more movie titles available in the VHS format.

Most servicepeople seem to feel that the VHS format provides a little more stability in the video output, with less flutter and "flagging" on the screen. Many also say that the VHS format is easier on the tapes.

On the Beta side of the argument, the initial quality of reproduction seems to be higher than with the VHS. And Beta's simpler threading and fewer tape guides mean that the machines are easier to service and repair in cases of a major malfunction.

Before you choose the format, find out which machine has the most readily available parts and service in your area. In general, the makes and models that are most popular in your area will be the ones that have parts the most readily available.

TOP LOADER OR FRONT LOADER?

Front-loading machines are gaining in popularity for two reasons. One is that it takes less effort on the part of the owner: Put the tape into the slot, give it a tiny shove, and in it goes. A top loader requires that you manually slide the cassette into the slot, and then manually push down on the drawer.

The second reason is that the front loader tends to do a better job of keeping dust out. A top-loading machine can be left open, and often is. The front-loading machine has a spring-loaded door that automatically closes whenever a tape is removed and right after one is loaded in. The inside of the machine is thus exposed for a shorter period of time.

The front loader does have a couple of disadvantages. Go to almost any store that rents tapes and VCRs, and you'll find that the vast majority use only the top-loading variety. There are two main reasons why this is so. First, a top-loading machine is generally less expensive. The explanation for that brings up reason number two—it's less expensive because it is simpler. This means in turn that there is less to go wrong.

A front-loading machine has to have a loading mechanism to grab the cassette and to pull it into the chamber (Fig. 3-1). That makes one more mechanism for which you have to pay, and one more that can malfunction. And, since it is electrically operated, if the power goes out, it could take some doing to extract the tape. (This same situation exists with those top loaders that use electricity to trigger the lifting mechanism.)

Some experts always recommend buying the least complicated machine that fulfills your requirements. This is generally the top loader; however, the front-loading mechanism rarely causes a problem and provides a number of advantages. There are no other appreciable differences between the two types of VCR. The servicing, maintenance, repairs, and other topics covered in this manual apply equally to both.

FIG. 3–1 The loading mechanism and its motor on a typical front-loading VCR.

BUYING A VCR

Once you have decided which format is best for your purposes, you are then confronted with hundreds of VCR manufacturers. You can select a VCR from one of the major companies, such as RCA, Sony,

Hitachi, JVC, Panasonic, Zenith, or Motorola, or you can buy a department-store "house brand." All VCRs are similar, but there are differences in control functions, playback/recording speeds, and the number of options available.

The unit can be as simple or as complex as you wish. The price of the VCR will vary depending on the number of options that you want. It's fairly easy to find a basic VCR on sale for $300 or less, and it's even easier to spend $1,500 or more. Options include clocks, timers, remote controls, and multiple heads, and you can even buy portable units with separate components.

If you are a fan of video music shows and are entranced by the new stereo videos, you may want to look for a VCR model that has stereophonic sound capability. The outputs of stereophonic VCRs are fed both to your television set and to your stereo sound system. A stereo-unit hookup is slightly different from the basic connections described in Chapter 5. The video output of stereo VCRs is fed to the antenna input connection of the television set, while the sound output is fed to the left and right channel inputs of your stereo system. A stereo hookup usually has two pairs of audio cables that are terminated with RCA pin or phono jacks. The owner's manual that comes with the stereo VCR fully describes the standard connections to both the television set and the stereo.

In selecting a VCR, also consider outside factors and the potential evolution of VCR technology. For example, a VCR can be used in conjunction with home personal computers. Home computers can use either a cassette audio tape recorder or a video tape to store information. There are even ways to link a computer and a VCR for playing games.

Two features that are enchancing the popularity of VCRs in this country are portability and adaptability. Many peripherals are available, such as cameras, switching devices, and editing machines. All these make it possible for you to make your own video programs. (The use of peripheral equipment and the techniques for switching, editing, and other forms of programmed production will be described in later chapters.) Portability allows you to take the VCR and camera with you, whether it's to make a video recording of an event or to make a home movie.

Videotape is rapidly replacing movie film for home entertainment. The initial investment in an 8mm sound movie setup is a little lower than the initial investment in portable VCR equipment, but between the cost of the film and the developing, you will have paid about $7 or more for only three minutes of film. With the same $7 you can buy a standard T-120 video cassette that can store up to six hours of recorded fun. To equal the same six hours of viewing, you'd need about 120 three-minute spools of 8mm film—and more if you plan to do any editing. At a cost of about $7 per spool, this comes to $840. Then there's the time and hassle involved in splicing the spools together. And while movie film takes a day or two to be developed, you can take the videotape home and watch it immediately. Another advantage: If you are not happy with a particular tape, you can record over it.

If you plan to enlarge your home-recording system in the future, you'll want a VCR that allows you to expand and adapt. It should probably be portable and have as many different types of inputs and outputs as possible.

Before you buy a VCR, have the salesperson or technician at the store show you how to "unbutton" the unit and what tools, if any, to use. Explain that you intend to perform your own routine cleaning and maintenance and therefore need easy access to the heads, tape guides, and other parts. On some models, the cover is held in place with Phillips screws. Other manufacturers use Allen, hex-nut, or standard slot-head screws.

Don't let the salesperson talk you out of doing your own maintenance by claiming that all you need is a cleaning cartridge. The salesperson may tell you that home-servicing is impossible and that maintenance and repairs can be performed only by qualified service personnel. But you know better. However, keep in mind that if you open the machine, even for routine cleaning, you will probably void the warranty. For this reason, don't attempt any servicing or maintenance until the warranty expires. If any malfunction occurs while the machine is under warranty, return it to the shop where it was purchased for repair.

Before buying a unit, new or used, talk to friends who own VCRs, or ask the salesperson for names of prior customers to determine if they are satisfied with their VCR.

NEW OR USED?

There are advantages in buying a new VCR, not the least of which is the warranty. There is also the satisfaction of pulling the unit out of its box and being the first person to use it. Of course, you pay for these privileges with a higher initial cost.

Buying a used machine can be a bargain, or it can be nothing more than a waste of money. It all depends on how you approach it.

Look through the newspaper ads and you'll find plenty of used machines for sale. The owner may be selling it because something has gone wrong with it or because he or she wants more features or a different format or perhaps has just grown tired of it.

This manual can help you find the bargain and avoid the lemons. A used machine selling for a very low price may have nothing wrong with it that a careful cleaning and adjustment won't fix. Many users ignore their VCRs until something goes wrong. Often the quality of reproduction has severely diminished simply because the machine is dirty. Few people would trade in a new car simply because it needs to be washed, but many do it all the time with VCRs.

When buying used equipment, take the same steps you would take if you were buying a used car. Take the unit for a "test drive," preferably in your own home with your own television set. If possible, have a qualified expert check it over. You will have to pay for this service, but it's better than risking the loss of several hundred dollars.

As with everything else, the simpler the machine, the less likely it is that something will go wrong. The more complex the machine, the more things there are to break. On a basic VCR that does nothing but record and play back—one that has no timers, no remote control, no fancy options—maintenance is easy. (Often, you can achieve the benefits of these options simply by using external devices. For example, you can use a simple timer to have the machine turn itself on and off to record a program.)

On the new units that have digital clocks and computerlike programming controls, there are many servomechanisms that are not directly involved with playback or recording. However, if any of these servomechanisms malfunction, they can render the entire unit inoperable. Servicing these devices is too complex for the home-handyman, and the unit will probably have to be taken to a qualified service repair shop for repair.

Many older VCR models were manufactured with the capability of picking up only VHF television signals (channels 2 to 13). If you buy one of these, you won't be able to record any program that is broadcast on UHF channels (14 to 83). Some models have limited capability when hooked up to a cable television system. Check which channels the VCR is capable of recording in order to decide if the "bargain" is worth it.

Another cautionary note applies to the channel-selection device. Some units have rotary channel selectors for both VHF and UHF stations, and some have push-button selectors. Be sure to find out what the procedure is for presetting the push buttons, and what the limitations are. For example, there may not be enough push buttons for the number of channels you want to tape.

LEGAL TAPING

There is one basic caution about copying recorded programs. Almost all commercial programs that are broadcast over public airways and those recorded on video cassettes are under copyright protection. Thus it is illegal to copy them, whether to sell to another party or to use privately.

To protect copyrighted material, many prerecorded cassettes have a *copyguard signal track* embedded directly on the tape. This is a type of scrambling code that allows the tape to be played on your VCR but prevents you from copying the program from one machine to another. (One of the most common methods of putting in a copyguard is to make the recording with a low-level horizontal synch pulse.)

Several of the laws governing copyright protection are currently under review as a result of the growing popularity of VCRs and the ease with which recordings can be made. Updates can usually be found in video recording and popular electronics magazines.

Copyright infringement is a serious offense. Be sure you know where you stand legally.

DUBBING

Dubbing is the transfer of a recording from one tape to another. To do it you'll need at least two VCRs—one to play the original tape and another to record. You can also place any number of other machines, such as editors and enhancers, in between. Of course, you'll need the proper cables and connectors to do this.

Dubbing is useful if you are editing a movie you have made, since you can rearrange the fragmented scenes in any way you wish. You can also make copies of recordings—for example, if you record your child's birthday party and want to keep a copy for yourself and send one to a relative.

Even though it is impossible to play a Beta format cassette on a VHS machine, it *is* possible to dub a program from one format to another. For example, a Beta cassette may be played on a Beta machine with the output of the Beta unit fed to the input of a VHS recorder. The VHS machine will make a copy of the original in its own VHS format. Almost all stores that sell VCRs carry connecting cables to hook the two machines together for dubbing.

SUMMARY

There are two basic VCR formats available—VHS and Beta. Both have advantages and disadvantages, but either will do a good job for you.

VHS tends to be easier on the tapes, which helps extend the life of the tapes and increase stability over a longer period. VHS machines are also more common, which makes purchase or rental of prerecorded material easier.

Beta machines produce a sharper image initially. The tape transport mechanism is less complex than that of a VHS machine, which means that maintenance is easier and repairs generally less expensive.

You cannot use a Beta cassette in a VHS machine, or vice versa. However, you can dub from one format to the other simply by using cables to connect the two machines. But keep in mind that some kinds of dubbing represent copyright infringement, which is a serious offense.

Buying a new VCR will give you a warranty, and you will be able to choose exactly which features you want. Buying a used VCR can save you money, but you must exercise a little extra caution before you buy. You don't want to end up with someone else's headaches. Chances are good that you can put a malfunctioning machine back into working order. But it doesn't make much sense to buy a used machine to save $150, and then spend $250 to have it repaired.

Chapter 4
How a VCR Works

As complicated as it seems, the principles of operation of a VCR are fairly easy to understand. Much of it is the same as with an audiotape recorder. An electromagnet imposes a specific pattern on the magnetic coating on the tape. The playback converts this pattern into the same voltage changes that created the pattern, giving you an identical copy of the original.

A BRIEF HISTORY

One of the great advances in modern home entertainment has involved improved techniques for recording sound and moving pictures. First came the still camera, then genius Thomas Edison's famed "talking machine." Sound recordings were originally made on wire cylinders, but things soon progressed to wax cylinders and then to plastic disks. Eventually, audiotape was developed.

Over the years, photography advanced similarly—from massive metal and glass plates requiring deadly chemicals, to roll film, to flickering motion pictures, and finally to modern methods whereby moving images may be permanently stored on magnetic tape. Combine the audio and video recording and you have a VCR.

An audio recording device converts sound waves into pulses of electrical energy and stores them. Playback is merely the reverse: Electrical pulses are converted into sound waves. A tape recorder does the same thing through the magnetized tape coating.

Eventually it was discovered that images of moving objects could also be stored magnetically. The image is scanned at exceedingly high speeds. The varying light intensity and colors picked up by this scanning process can be converted to electrical pulses, stored, and played back at a later time.

The first videotape recorders were huge and tremendously expensive, usually the exclusive property of profitable television stations. The initial prototype video recorders made available in the 1950s cost in excess of $100,000 each. They were reel-to-reel devices, with plastic-based videotape that rolled from one reel across a recording or playback "head" to the second pickup reel. To be able to handle all of the electrical impulses involved in storing the sound and video images, the early tape was two inches wide and was moved across the heads at high speed.

By the 1960s enough advances had been made in reel-to-reel machines to make it possible for the average person to own one. The problem was that simple black-and-white machines cost in the $6,000-plus range. They also had the unfortunate tendency to break down, and there were few technicians around who could repair them.

TAPE LOADING

A great improvement in the convenience of sound recorders came with the advent of tape cassettes. These were like miniature versions of reel-to-reel, but with the two reels attached and safely enclosed. A video cassette is somewhat similar to an audio cassette: Both eliminate the need for tedious threading of the tape through the machine, and both protect the tape, which is nestled inside the casing.

With this design, the tape is drawn from the cassette, fed through the transport mechanism, and taken up on the opposite spool. All this is done automatically, without your ever touching the tape. When a cassette is placed into the player and "Play" or "Record" is activated, several motors begin spinning, driving the capstans (round metal dowels), the supply and takeup spools, and the main video head.

A rubber pressure roller (pinch roller) is mechanically pushed into position, holding the tape firmly against the spinning capstan. The tape is literally squeezed between the capstan and pressure roller. It is then pulled from the supply reel and fed through a series of tape guides, which direct the tape across the heads (see Figure 4–1).

The tape is then drawn across the recording or playback heads. There are usually at least three heads, two of which are not mentioned in the advertising. First comes the erase head. Its function is to erase whatever is on the tape, making it possible to record over an existing recording. This head is active only during "Record." The main video head is a drumlike assembly, shown in Figure 4–2. The actual head is

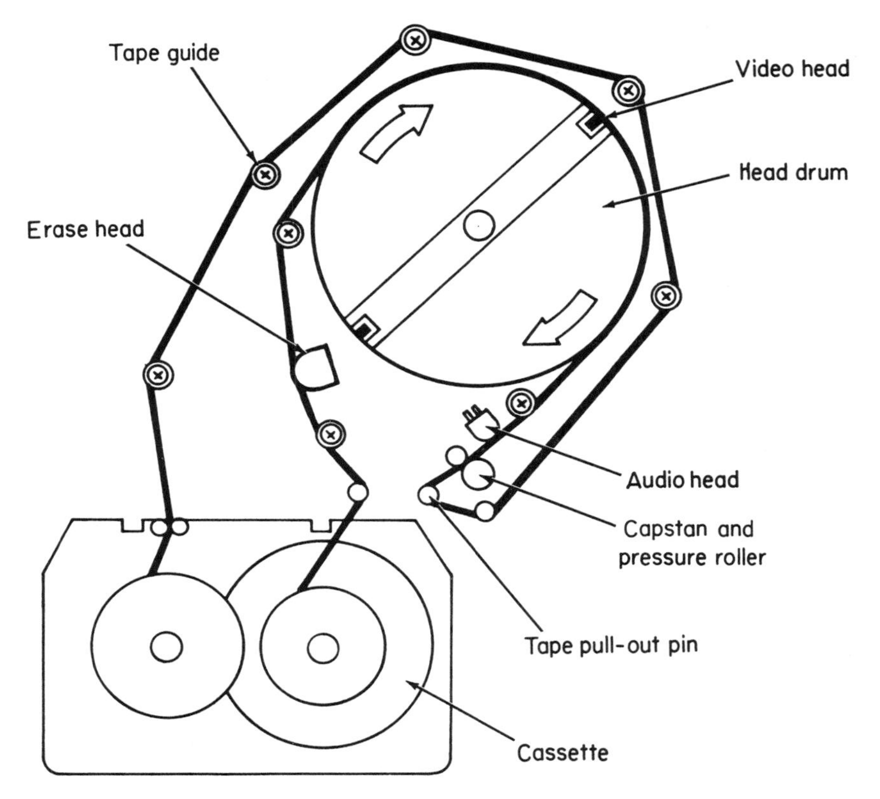

FIG. 4–1 Tape threading path.
Reprinted by permission from Harry Kybett, How to Use Video Tape Recorders *(Indianapolis: Howard W. Sams & Co., Inc., 1974), p. 145.*

inside the drum and spins at about 3,600 rpm. The tape passes over the audio head. Finally, the tape is taken up on the other reel. (For more information on tape loading, see Chapter 6.) This complicated operation is done automatically. All you have to do is press a button.

As with an audiotape, a videotape is divided into several *tracks.* A track is merely the path along the tape that is traced as the tape

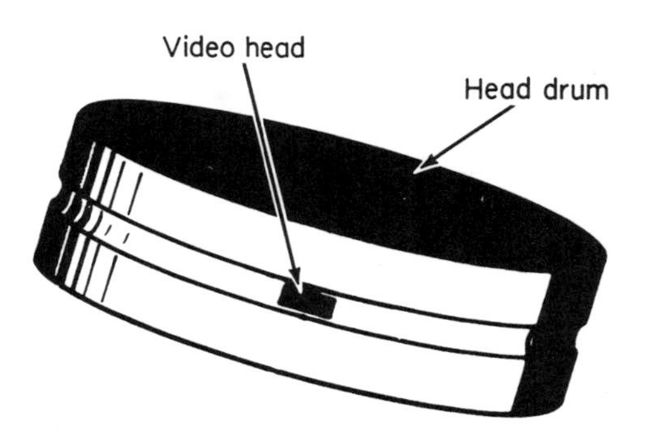

FIG. 4–2 Head drum assembly. *Reprinted by permission from Harry Kybett,* How to Use Video Tape Recorders *(Indianapolis: Howard W. Sams & Co., Inc., 1974), p. 187.*

passes over the head (Figure 4–3). If a manufacturer has designed a $\frac{1}{4}$-inch-wide reel of tape to be a four-track audio cassette, then the recording and playback heads necessary for properly storing and reproducing the sound must be no more than a sixteenth of an inch wide. When the tape crosses the head to play the first track, the head is positioned so that the bottom sixteenth inch of the tape touches the head. To play track two, the head is moved up to the second sixteenth of an inch, and so on.

Equipment today has been miniaturized so successfully that a narrow $\frac{1}{4}$-inch tape can handle as many as 32 separate tracks. It's easy to see how critical it is to maintain a smooth and tight motion across the fixed-position heads, with no flutter or slippage as the tape rubs across the delicate head assembly.

The tracks of an audiotape are linear, running in a straight line from one end of the tape to the other, whereas the tracks of a video-

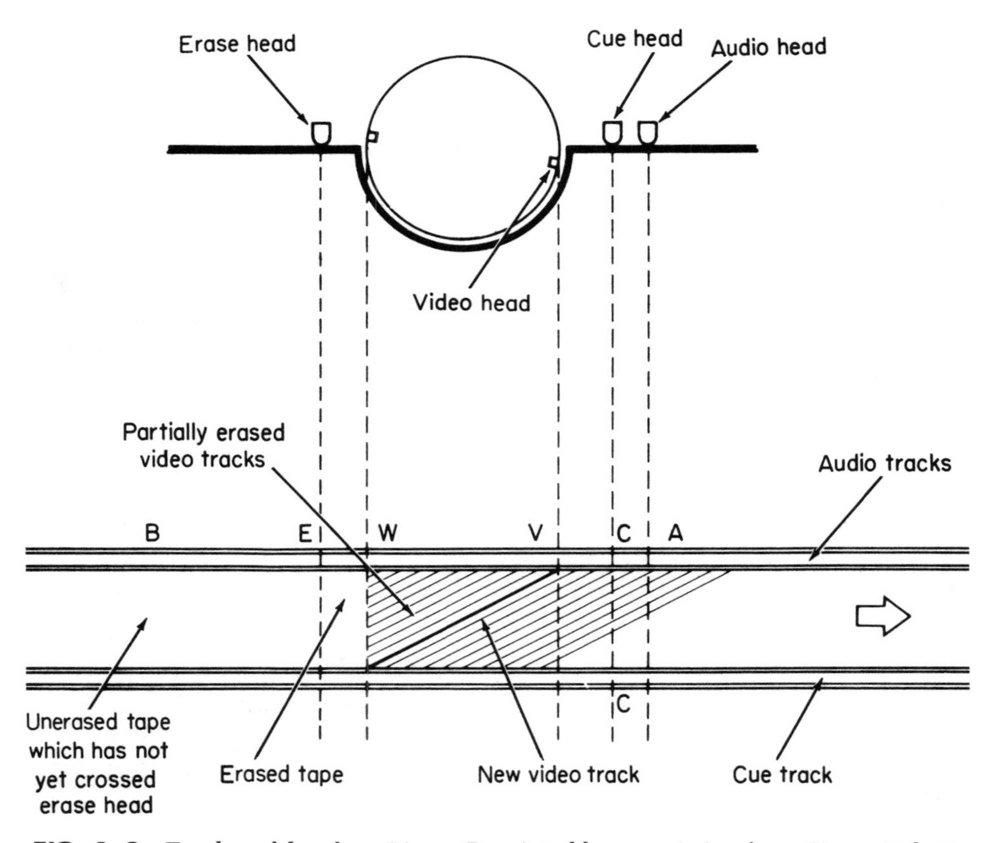

FIG. 4–3 Track and head positions. *Reprinted by permission from Harry Kybett,* How to Use Video Tape Recorders *(Indianapolis: Howard W. Sams & Co., Inc., 1974), p. 83.*

tape run diagonally across the tape. The difference is due to the way the tape rides on the drum.

To achieve quality reproduction, high-speed transport is required. This can be achieved simply by feeding the tape through the machine at high speed, but this obviously requires a lot of tape and a cassette too big to be convenient. Further, stabilizing a two-inch-wide strip of tape inside a cartridge is difficult at best. Not only does the tape width have to be fairly narrow, but the overall length has to be short enough to fit inside a reasonably sized cartridge. This means that the tape speed has to be fairly slow. Yet, to scan the video section effectively, the tape has to move fast.

A new design was developed to make it possible: A narrow tape short enough to fit inside the cartridge. To provide the necessary scanning speed, the record and playback heads were made to spin crosswise to the tape at a fairly high and constant rate.

TAPE SPEEDS

Most VHS units have three tape speeds: SP (Standard Play), which provides two hours of recording or playback; LP (Long Play), which provides four hours; and SLP (Super Long Play) or EP (Extra Play), which gives six hours of play. Beta machines have only two speeds—SP and EP. (EP is sometimes called LP on Beta machines.)

Basically, the faster the speed, the higher the quality of reproduction. You may hear that a VCR isn't quite as reliable at the slowest speed. This isn't true. The quality will suffer somewhat, but the tape will feed through properly at any speed.

THE CAMERA

A scene to be taped or televised is scanned by a video camera in much the same way as your eyes would scan the page of a book. The eye begins at the upper left-hand corner of the page, reads the first word, and moves across the line to the last word. Then the eye instantly flicks back to the left edge of the second line and again moves horizontally across the page. The process is repeated from top to bottom on the page.

When the electronic circuits in a video camera scan an image, a light-sensitive "eye" moves across the scene from upper left to upper right. As this horizontal path is traced, the variance in brightness, or light level, is converted by the moving electronic "eye" into a voltage that varies exactly in step with the intensity of the light. Different colors produce certain voltage levels.

When the "eye" reaches the upper right-hand corner of the scene, it "blinks" for a fraction of a second while it moves back to the left corner of the scene and slightly downward so that it may then trace a second horizontal scan just below the first trace.

While it may take your eye a minute or two to read a complete page, this electrical scanning by the camera is almost instantaneous. A standard television picture is reproduced through the tracing of 525 horizontal lines of varying light level from the top to the bottom of the scene. These 525 lines are scanned from left to right, top to bottom, thirty times each second. (Get out a calculator and you'll find that this translates to nearly 160,000 traces per second.) At this speed, the eye simply cannot follow the individual lines as they are being traced ("written") across and down the screen. Instead of flickering light, we see a reproduction of the original scene as the varying voltages are painted back on the television screen in the form of changing light levels.

THE MICROPHONE

Sound waves are picked up by a microphone, inside of which is a tightly stretched diaphragm. The sound waves cause this to vibrate. In one type of electronic conversion, the diaphragm is attached to a coil of wire. As the diaphragm moves, the coil of wire is forced to move through the field of a powerful magnet. (The movement of a coil of wire is a magnetic field is a common way of generating electricity.) The electrical voltage generated by the moving diaphragm in a microphone is very low, but it can be amplified, just as the video signal is amplified.

THE TELEVISION SET

By the time the signals travel through the air from the television station to your antenna, they are once again very low in voltage. The amplifiers of the television set take over and boost the signals to a level that can drive the speaker and picture tube. But despite the raising, lowering, and raising again of the voltage level, everything is kept in step with the original. The video voltage "repaints" the original scene on the television screen; the sound voltage re-creates the original voices and sounds in the set's speakers.

Cable television operates in the same way, but with cables bringing the signals directly to your home and television set. No antenna is needed, and atmospheric distortion is minimized. Recording and playback of home video and sound are similar, except that there isn't air or a cable system (other than your own connections) stuck in between.

HOME RECORDING TAPE

Recording tape begins as a long roll of plastic and is specially treated. One of the most popular plastics for this use is polyethylene terephthalate, or simply PET. A brand name for PET is Mylar, owned by DuPont.

One side of the tape is coated with billions of tiny molecules of magnetic material, usually ferric oxide. The tape is like a conveyor belt covered with billions of tiny bar magnets, each with a north and south magnetic pole. The incoming signal to be recorded causes these molecule-sized magnets to take on the magnetic charge.

The basic rule of magnetism is, "Like poles repel; unlike poles attract." This causes the now-magnetized particles on the tape to align themselves in the same direction as the voltage fluctuation passing from the amplifiers and along the wire to the recording head.

The voltage applied to the head changes in step with the sound and picture information originally scanned. On a VCR, the heads spin constantly at a standard 3,600 rpm. As the tape passes across the recording head, the magnetic molecules lock into new alignments, which correspond to the changing voltages applied to that head. The images and sounds produced are now stored on the tape in the form of magnetic "words."

After the tape has been rewound, it can be played back, with the VCR's machinery operating in reverse. As the tape moves across the heads, magnetic alignments on the tape activate the playback head. This generates varying voltages that exactly duplicate the images and sound recorded onto the tape. The varying voltages are fed to the amplifiers of the VCR and your television set. The sound voltages are used to drive the cone of the speaker back and forth, which generates sound waves. The video voltages are sent to the picture tube, which swiftly "repaints" the horizontal light tracings of the original image in 525 lines retraced 30 times per second—just as they were originally scanned by the camera.

As simply as this, a recorded scene is replayed on your home video screen.

SUMMARY

A video camera scans a scene electronically, measuring and recording differences in brightness and tone. Meanwhile, the microphone converts sound waves into electrical signals. Both signals are fed to the recording heads, and the varying voltages cause a change in the magnetic field. This pattern is stored on the tape by charging the tiny magnetic particles that coat the tape. Later, this pattern of charge can be picked up by the playback heads and converted back into an electrical pattern that exactly matches that of the original.

These variations in voltage can be sent through cable or through the air to your television antenna. However they are transmitted to your home, the principle is the same. So is the principle behind making your own recordings. The incoming signal causes a voltage change on the heads, which creates a magnetic pattern on the tape.

An audiotape signal, less complex than a video signal, moves in a straight line across the heads. A videotape records the video image on diagonal tracks.

Loading of a VCR cassette is automatic, occurring either when you insert the cassette or when you push the "Play" button. Capstans, pinch rollers, and pins pull the tape from the cassette, run it through the tape guides, and wrap it around the heads. The heads then "read" the magnetic signals stored in the particles—and you have a beautiful reproduction.

Chapter 5

Making the Connections

Before operating your new VCR, study the owner's manual provided with the unit to learn the location and use of all of the control switches, functions, and external connectors.

Most VCRs today have similar basic controls: a power switch, play control, record switch, stop switch, fast forward, rewind, and cassette-eject controls. Depending on the sophistication of the unit, a VCR may also have a digital clock display, a timer control to preset recording schedules, channel selectors (either rotary or push-button), pause control, single-frame advance, tracking control, a speed switch, and a counter to measure how much tape has passed across the heads.

All VCRs operate through channel 3 or 4 of your television set, depending on which of these channels is not in regular use in your area. There will be external connectors for the following items (see Figures 5-1 and 5-2):

1. VHF (and sometimes UHF) *inputs* from the antenna to the VCR.
2. VHF (and sometimes UHF) *outputs* from the VCR to the television set that is to be used as a monitor.
3. Audio output/input, which allows you to feed the sound signal from your VCR to a separate speaker or monitor or to another VCR. The audio input permits you to bring sound from a microphone into your VCR for recording. A separate audio hookup is not necessary for normal recording and playback of regular television shows. The VHF inputs and outputs (items 1 and 2) provide for both picture and sound information of all television recording and playback performed through your television set.

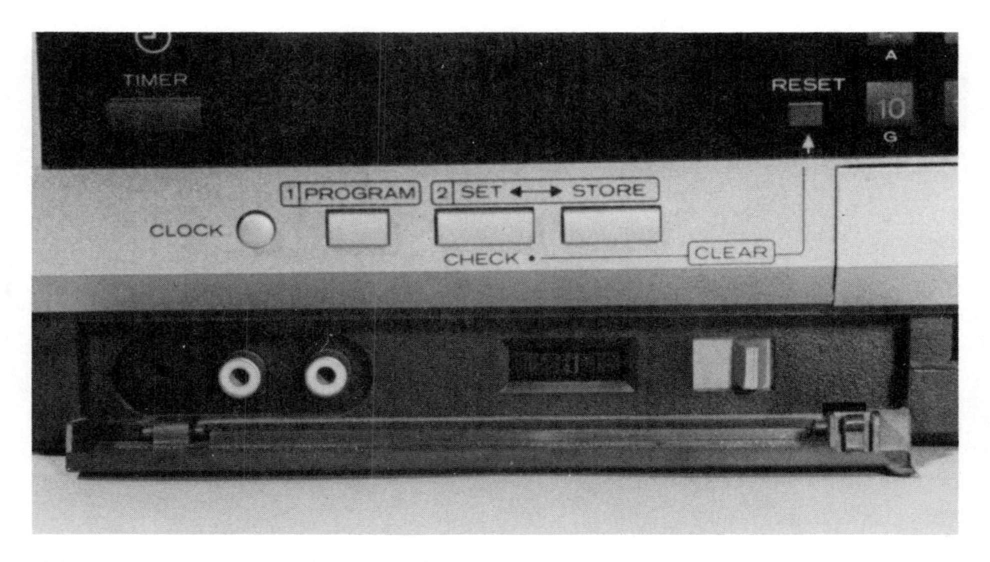

FIG. 5–1 Inputs and outputs on the front panel.

4. Video output, which allows some type of connection from your unit to a second VCR for recording from one unit to another, or for editing.
5. Camera input, which is designed to accept the output of a camera from the same manufacturer as the VCR, or of a compatible type.

FIG. 5–2 Inputs and outputs on the rear panel.

There may also be an external connection for a remote-control device sold by the manufacturer in conjunction with the VCR. And many VCRs come with a separate earphone jack to permit private listening.

If you intend to move your VCR frequently, keep your connections as simple as possible. Push-on connectors can be used for a quick connect/disconnect to give you maximum portability. Make sketches of your wiring scheme so you'll know how to connect everything again. (You may even want to label the various cables.)

CONNECTORS AND CABLES

Adaptors, connector cables, and matching transformers for the standard VCR-to-television hookup are usually supplied with the VCR (Figure 5–3). The number of connectors you will need for your own hookup depends on what you have in mind and also on whether your home is wired for cable-television pickup.

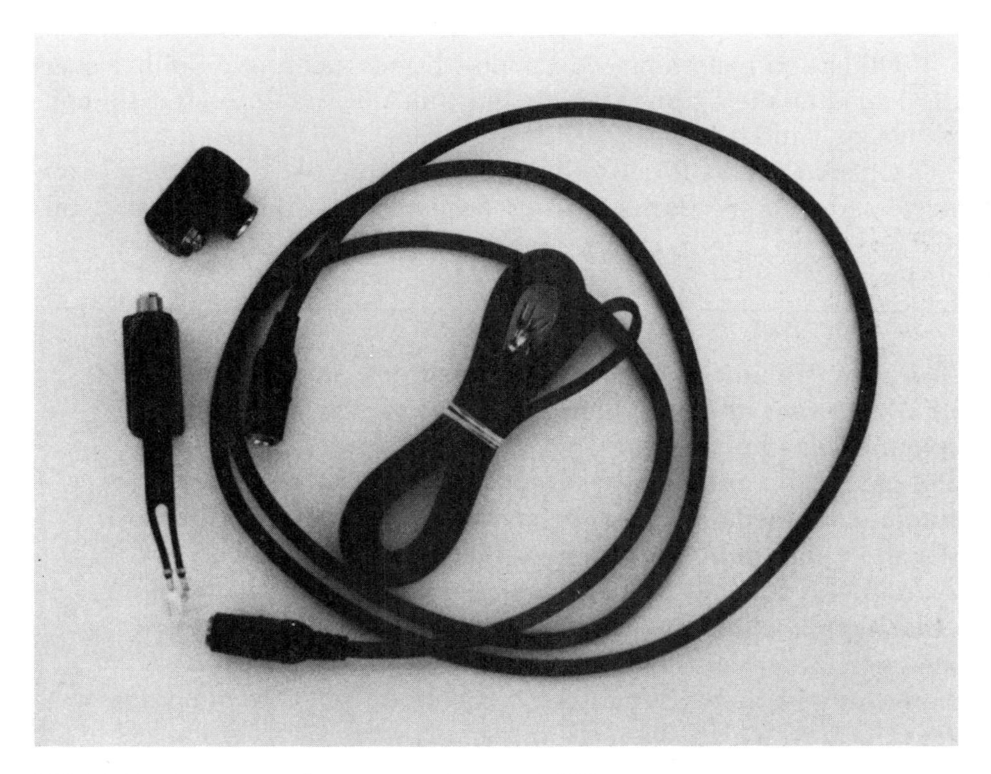

FIG. 5–3 Connectors that come with a new VCR.

The basic items usually needed include:

1. A five- to ten-foot length of 75-ohm coaxial connector cable. The end of the cable should be fitted with a $\frac{3}{8}$-inch RF male screw-on or slip-on connector. (Some VCRs use a $\frac{1}{4}$-inch RF coaxial cable. Be sure to find out which one is correct for your unit.)
2. A 300-ohm-to-75-ohm VHF adaptor. This adaptor accepts the 300-ohm impedance of a twin-lead antenna wire and changes it to a 75-ohm impedance for the television set or VCR, terminated with a male RF connector.
3. A 75-ohm-to-300-ohm matching transformer with a male $\frac{3}{8}$-inch RF connector at one end and a short section of 300-ohm (flat) twin-lead antenna wire on the other end, terminated with two spade lugs. This adaptor does exactly the opposite of the one described in item 2: it takes a 75-ohm input, such as from a cable, and changes it to a 300-ohm output for the television set.
4. A five- to ten-foot section of 300-ohm (flat) twin-lead antenna wire to connect the VCR to the television set.

Other connectors provided with the set incorporate either standard or miniature phone plugs, the $\frac{3}{8}$-inch RF male or female couplings, or standard RCA pin jacks.

Check your VCR carefully to see which types of plugs and connectors you should use. Then purchase one or two replacement con-

The ohm is a unit of impedance measurement. A 75-ohm cable offers 75 ohms of impedance to the flow of current along its length. This impedance is 75 ohms regardless of the length of the cable. You must match impedances whenever you connect audio and video devices to each other. Improper matches will result in poor quality or in no signal transfer at all. For example, if your VCR has a 75-ohm output connection, you cannot connect it to a 300-ohm antenna without losing signal strength. The picture will be snowy and will lose detail, or there may be no picture at all. A 75-ohm-to-300-ohm matching transformer, which converts impedances from 75 ohms to 300 ohms, must be used between the devices.

nectors of each type for your toolbox stock. You can buy these connectors already attached to the proper length of cable from your VCR dealer, an electronics supply store such as Radio Shack, or the electronics department of most major department stores.

You can save money if you take the time to learn how to attach these connectors yourself. The only tools required, including a low-wattage soldering iron, are listed in Chapter 1.

SPADE LUGS

A spade lug (Figure 5-4) is a U-shaped connector designed to be slipped under the head of a screw and firmly attached to a terminal. The other end of the lug is either a small cylinder or fold-over leafs designed to wrap around a straight section of wire. Spade lugs are usually used at the ends of all 300-ohm (flat) twin-lead television-antenna wire.

To attach a spade lug, first strip the two ends of the twin-lead wire to expose about $\frac{3}{4}$ inch of bare wire (Figure 5-5). (Check to see that the wire is clean and not corroded. If it has turned green or gray, or any color other than the "healthy" golden shine of copper, cut off an inch or so of the wire and start again.) Twist the tiny strands together on each side, and use the soldering iron to tin (apply a small amount of solder to) the ends (Figure 5-6).

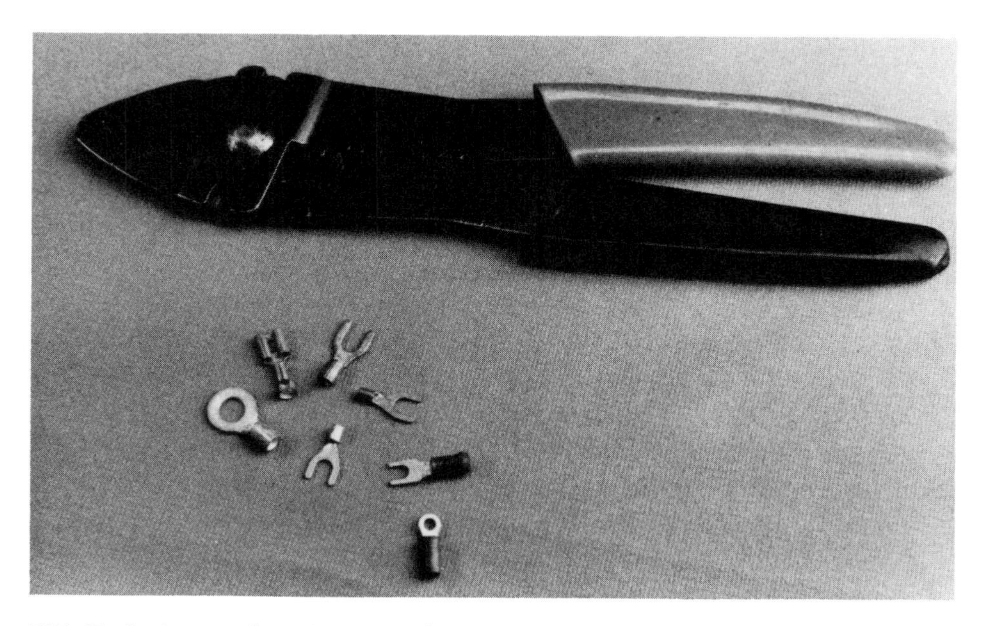

FIG. 5–4 Lugs and a crimping tool.

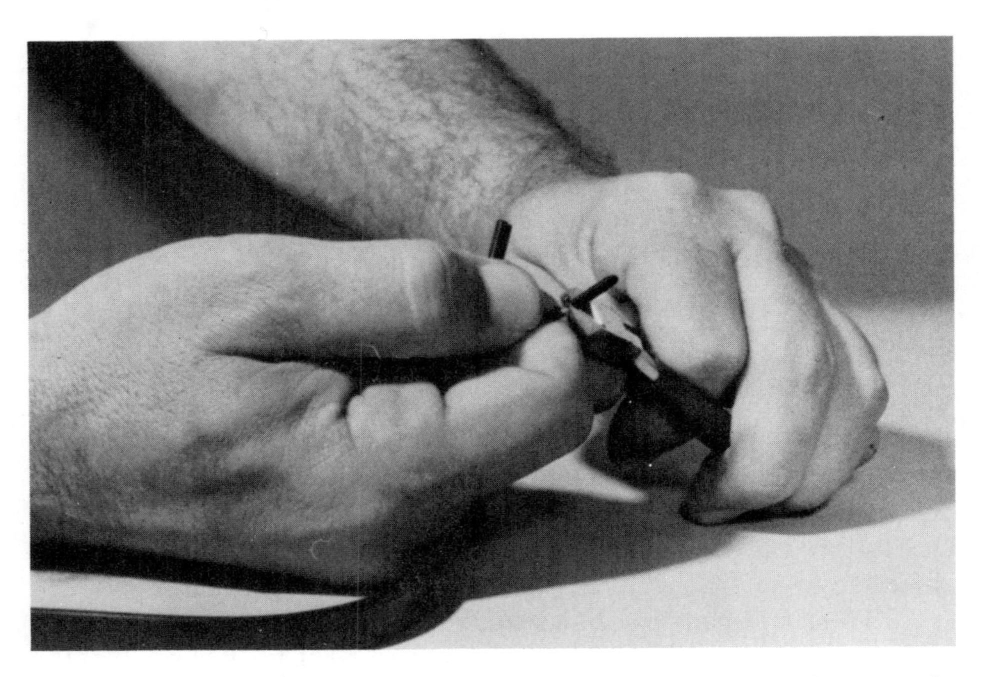

FIG. 5–5 Strip the ends so that about $\frac{3}{4}''$ of bare wire is exposed, then twist the wires.

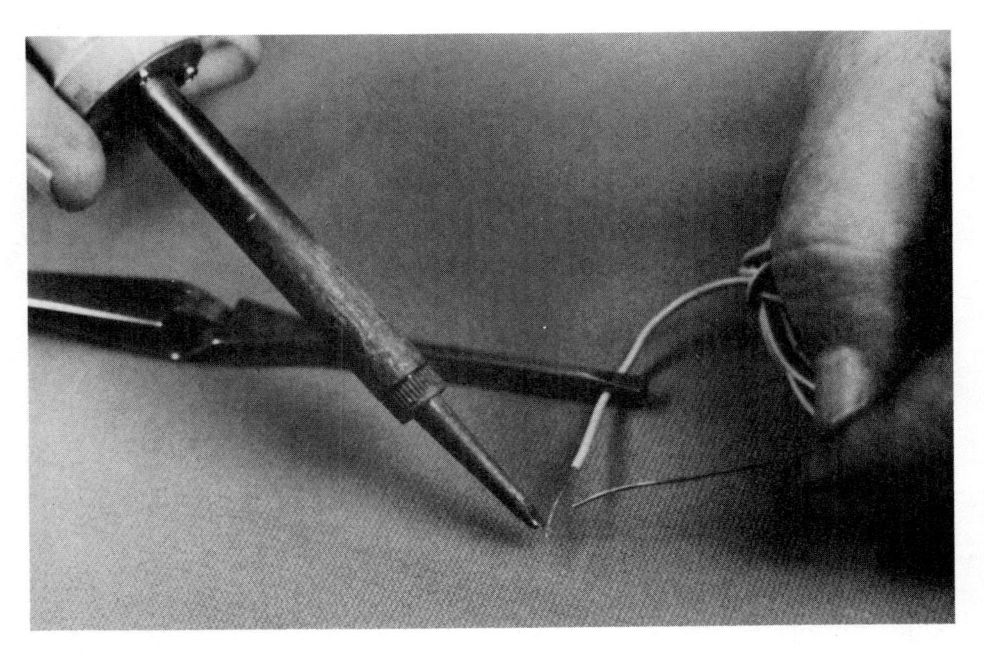

FIG. 5–6 Tin the twisted wires.

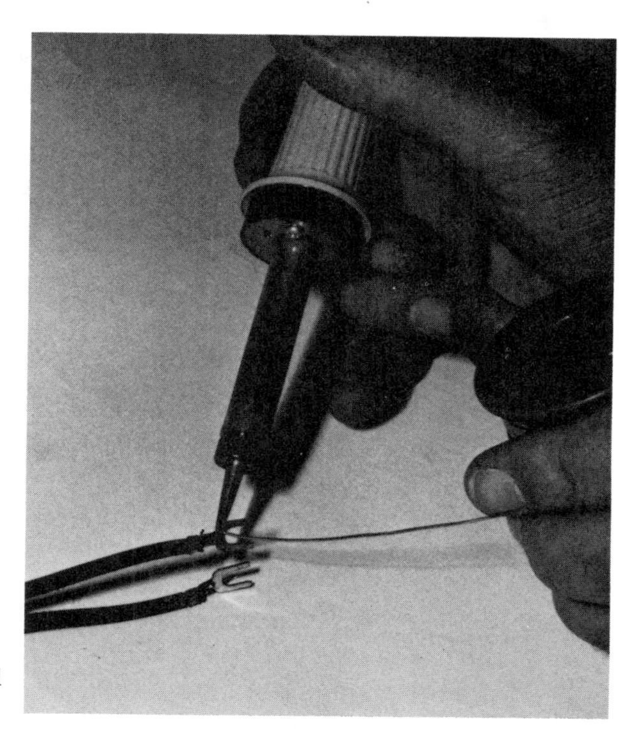

FIG. 5–7 Solder the tinned wire into the lug.

Using a small cutter or sharp knife, slice down the middle of the flat plastic insulation between the two wires. The cut should be about 3 inches long. Then very carefully cut away the excess plastic material. DO NOT cut into the insulation around the wire itself. If you do, clip off what you've already done and start over. (You may want to wait to tin the ends until after you've removed the insulation, just to save time in case you make a mistake.)

Attach each spade lug to the wire either by pushing the wire through the cylindrical end of the lug or by laying the wire between the wraparound leafs of the lugs. Either way, be sure that no bare wire is exposed beyond the lug. You may have to trim the end of the bare wire slightly to get a perfect, professional fit.

Crimp the holding metal tightly to the wire. Apply the tip of the soldering iron to this connection and carefully apply enough solder to make a firm electrical connection (Figure 5–7).

PHONE PLUGS

Phone plugs (Figure 5–8) come in standard $\frac{1}{4}$-inch, miniature $\frac{1}{8}$-inch, or subminiature $\frac{1}{16}$-inch diameter. It is best to purchase subminiature

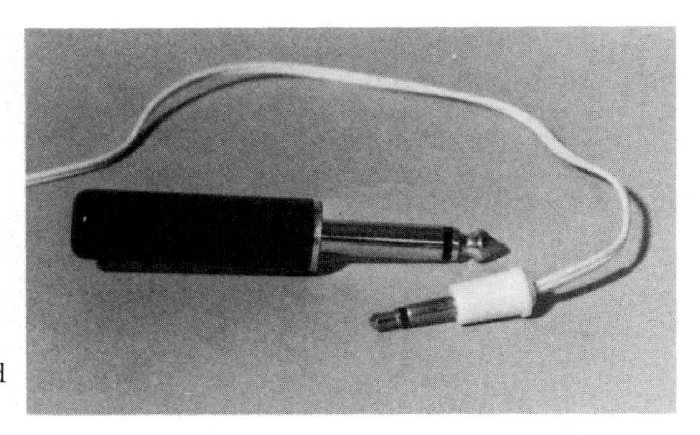

FIG. 5–8 Phone plug and miniature phone plug.

spares with the proper-length cable already attached, since their small size makes it difficult to build them at home.

Audio cable used with phone plugs is an insulated, two-wire conductor. Usually, there is a solid center wire coated with plastic or rubber insulation. This insulation is wrapped with a shielding lattice wire, which is covered by a second layer of insulation.

Care must be taken in stripping the ends to be sure that the two wires do not become exposed and touch (short out) at any point. Use wire strippers or a sharp knife to remove the outer insulation layer from the last $1\frac{1}{2}$ to 2 inches of the audio wire to expose the shield.

Carefully unwrap the shield. It is usually wound in one direction around the inner insulated wire. Twist the strands of exposed shield wire together and tin the tip. To do this, apply the tip of the soldering iron to the end of the exposed shield and lightly solder it for about $\frac{1}{2}$ inch of its length. Be careful not to let the tip of the soldering iron get too close to the remaining portion of insulated cable. The head could melt the insulation and cause a short, forcing you to cut off the end and begin all over again.

Occasionally, audio cable is insulated with a special RF shielding lattice. Four fine strands of wire are tightly wound around the insulation of the inner conductor along the full length of the cable. Then four more strands are wound in the opposite direction along the length of the cable. This winding of four strands on top of four strands is repeated, with the direction of winding alternated each time, until the inner cable is completely covered. This provides a very tough cable that will completely shield the inner conductor from stray RF (and block RF from escaping from the center conductor).

The easiest way to remove this complex shield from an inner conductor is to use the tip of a pointed object, such as a small nail, to pry a small opening in the lattice-work shield and expose the inner insulated conductor. Carefully work the point of the nail between the shield opening and the inner cable until you are able to nurse the inside conductor through the hole. The nail or a screwdriver can help. Pull the inner cable out enough to form a loop through which you can put the nail or screwdriver. Then just pull the inner conductor out.

After separating the inner conductor from the shielding, tin the outer shielding and strip away about $\frac{1}{4}$ inch of the inner insulation to expose that center conductor.

RF CABLE AND CONNECTORS

RF cable is stripped from the end for the connection of the male or female connector in the same manner as described for phone plugs. The shielding of RF cable is almost always of the lattice or interlaced-type winding. The outer insulation is a tough rubber coating, while the center connector is covered with a white paraffinlike substance. Because of the design of RF connectors (Figure 5–9), you will be able to cut away most of the exposed shield when you solder this cable to the proper conductor.

Once you have stripped the center wire, slip it inside the hole in the connector. The center wire actually becomes the male prong of the connector.

When purchasing connectors for the end of any RF cable, be sure to buy the correct size. RF connectors used with VCRs are usually either $\frac{1}{4}$ or $\frac{3}{8}$ inch in diameter, and they are *not* interchangeable.

Some connectors do not require solder. Instead, the inner conductor with its insulation and stripped "male prong" is slipped into a tight-fitting sleeve. The outer shield is pushed back along the outside

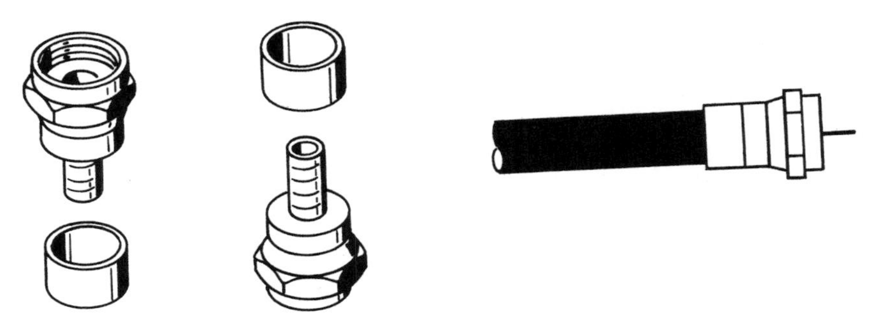

FIG. 5–9 Closeup of RF connectors.

of the rubber insulation, and clamps from the connector are firmly pressed around the shield with pliers or with a special crimping tool (Figure 5-10).

On other types of RF cable, the inner conductor is slipped through a tight-fitting hole and solder applied. The proper method of doing this is first to insert about $\frac{1}{2}$ inch of solder wire into the hole. Then use the tip of the soldering iron to heat the sleeve until the solder has fluxed and slip the tinned inner conductor tip into the presoldered and preheated sleeve. Finally, remove the soldering iron, holding the tip and cable rigid until the solder cools. If the cable or connector wiggles before the solder cools, you may have to reheat the joint and try again to get a good connection.

After firmly attaching the center conductor, thoroughly solder the outer shield to the outside holding clamps.

RCA-TYPE PLUGS

When you buy RCA pin jacks (Figure 5-11), you must specify whether you need chassis-mount or cable-end types. For VCR connections, you will almost always want the cable-end type. These have a sleeve hole, much like RF connectors, for the inner conductor. About $\frac{1}{2}$ inch of solder is slipped inside the sleeve, the sleeve is heated with the iron, and a tinned inner wire of your cable is slipped inside the sleeve.

The shield is usually soldered to a lug hole, and then either a plastic or metal outer case is screwed on the completed connection. Pin jacks are designed for use with audio cable. They are sometimes used as the video connector from an external camera to the VCR.

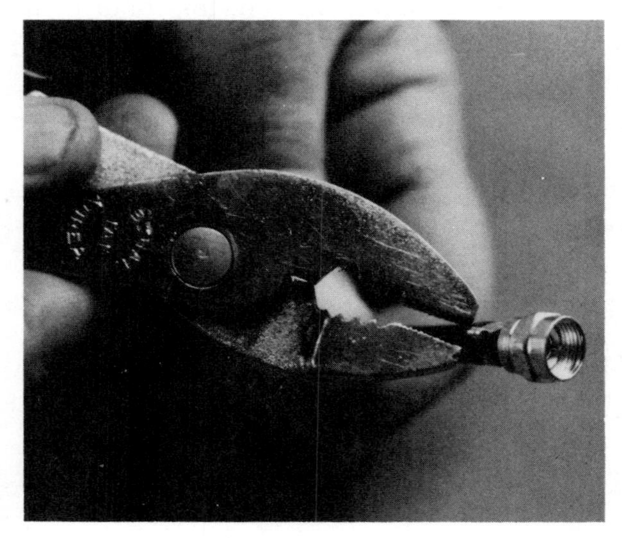

FIG. 5–10 Crimping a solderless RF connector.

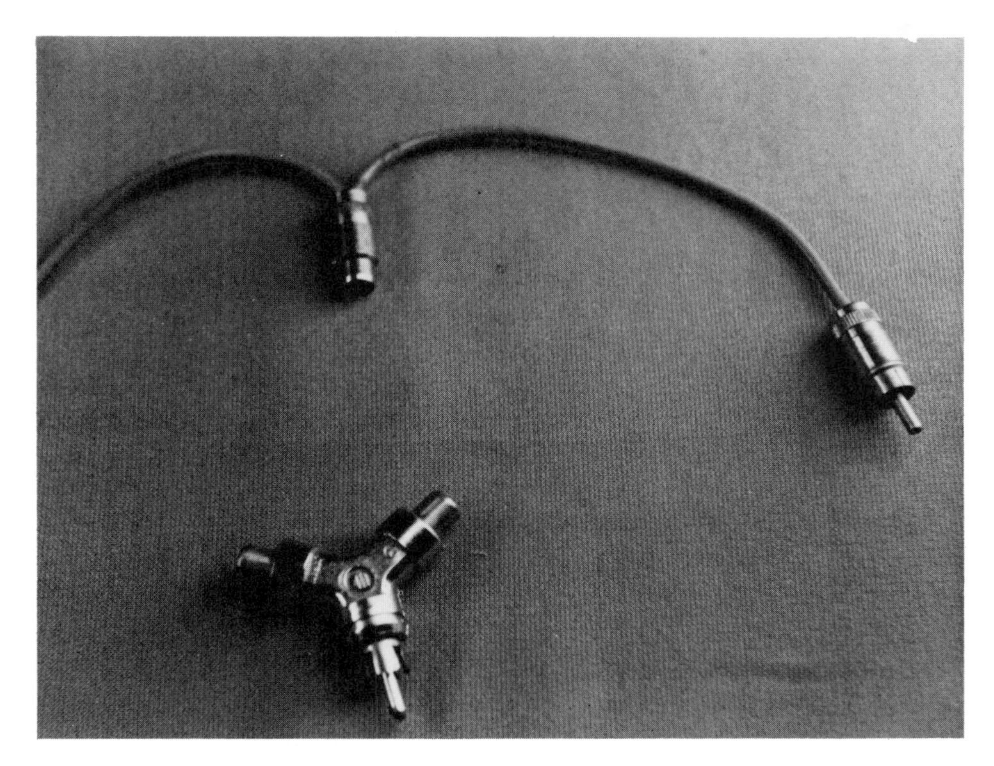

FIG. 5–11 Phono plugs, sometimes called pin jacks.

TESTING THE CABLE

After making any soldered connections, always check the final results with your VOM (Figure 5-12). Put the meter on a low-resistance scale to read continuity. Test the meter by touching the two probes together. The meter should deflect full scale to show zero resistance.

To test the continuity and integrity of a cable, use the two meter probes. If you've soldered a conductor to a pin or lug, touch the probe to that pin or lug, *not* to the conductor itself.

Now take the cable to which you have just attached a new conductor. Put one probe on the pin connected to the inner wire and touch the outer probe to the shield connection. The meter should show infinite resistance, which tells you that the inner and outer wires are not touching. If you get a reading other than infinity, the two conductors are touching somewhere, which means that the connection is poor.

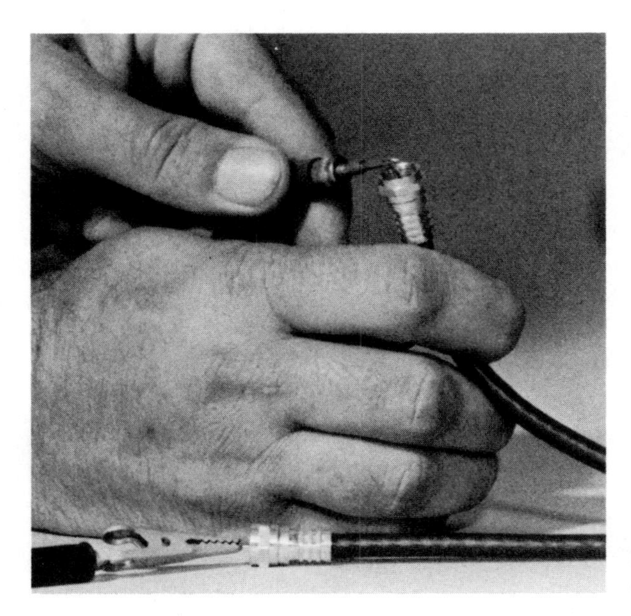

FIG. 5–12 Testing the cable for continuity.

Next, put one probe on the inner conductor at one end of the cable and the other probe to the inner connector at the other end of the cable. The meter should read zero ohms, signifying that there is a good connection.

Finally, put one probe on the outer shield at one end of the cable and the other probe to the outer shield at the other end of the cable. Again, the meter should show a full deflection—zero ohms—indicating a solid connection.

MAKING THE CONNECTIONS

The way television signals are received in your home determines how your VCR will hook up to your television set. Signals can be received from:

1. An outside antenna with either 300-ohm (flat) twin-lead antenna wire or 75-ohm coaxial cable running from the antenna to your television set.
2. "Rabbit ear" antenna(s) sitting on top of your television set.
3. A 75-ohm coaxial cable from a commercial cable firm's channel-selector box running to your television set, or from a satellite receiver.

Regardless of how television signals are received in your home, the VCR must be connected between the antenna and your monitoring television set.

An important thing to remember at this point is that your VCR is almost exactly like a small television set in itself, but without the picture tube. In other words, the VCR can work in its recording mode completely independent of your television set. As long as an incoming signal is going to the VCR, your television set has absolutely nothing to do with the signal going to the videotape through the VCR when the unit is recording.

The VCR has an antenna connection, a channel selector, devices for separating the picture and sound signals, and amplifiers—devices found on all television sets, whether black and white or color. The only television items that the VCR lacks are a picture tube and speaker.

In the playback mode, the VCR output is fed to the television set on either channel 3 or 4. If the selector controls on your set are turned to any other channel, you will see the signal from that channel.

When your television set is turned on for normal reception, or when your VCR is turned off or in the playback mode, the antenna connections are switched directly through the VCR to the television set. When a VCR is in the record mode, the antenna feeds a split signal simultaneously to the VCR and the television set. Your antenna, rabbit ears, or cable-television feed contains the signals of all television stations from your area at the same time. The channel selector on the VCR or the television set determines which particular channel you will be recording or viewing.

With the antenna input connected to your VCR and the VCR output connected to your television set, you can record one channel while watching another. For example, the VCR could be set to record a movie on channel 10 while you're watching a special on channel 5—or while the set is turned off and you are sound asleep.

You could turn the television set on or off or even unplug it and the VCR will still record whatever signal it is receiving from the VCR's own channel selector. But remember, your antenna must be attached to the VCR and the VCR connected to the television.

THE CHANNEL 3 OR 4 SELECTOR

Determine whether you want your VCR to come into your television set on channel 3 or channel 4. If a local station broadcasts on channel

3, use channel 4 for your VCR, or vice versa. If, in a rare instance, local stations broadcast on both channels, set your VCR to the *weaker* (farthest away) of the two. Ask your dealer which of the two works best in your area.

Your owner's manual will show you how to find the channel 3 or 4 selector. Make sure that the selector is switched to the same channel as the television set.

ANTENNA HOOKUPS

A single outside antenna can be used to feed up to *four* different televison sets in a home. That same antenna system can also feed the input of the VCR, with the VCR output feeding more than one television set, so that recorded programs can be watched simultaneously on more than one set in various rooms of the house.

Best viewing requires a top-quality antenna. The better the antenna, the stronger the signal will be. All of the cables, wires, and connectors used to connect the antenna to your various television sets and your VCR should be in good condition. If any show signs of wear or fraying, replace them—or expect a poor signal.

You can connect everything with 300-ohm (flat) twin-lead antenna wire. There are, however, some disadvantages to using this type of wire. Since the two wires run side by side, they are more prone to interference from outside sources. Stand-off insulators are needed to keep them from touching the house. Care must be taken to ensure that the wire has no sharp bends in it. The length must be exact, since coiling any excess can cause a poor signal.

For best installation, use 75-ohm coaxial cable for all connections. With coaxial cable, the signal is carried by a central conductor encased in a paraffinlike plastic. A braided-wire lattice surrounds this, shielding the center conductor from interference. Coaxial cable therefore doesn't need to be installed with the insulated stand-offs to keep it from touching the house as it comes down from the antenna, nor do you have to be concerned about sharp bends or coiling. You can run the cable right alongside electrical and telephone wires and other metal objects that would cause interference if you had used 300-ohm antenna wire.

Conversion is not difficult if your antenna and television sets are already hooked up with the old-fashioned 300-ohm antenna wire. If your antenna and the back of your television set are already equipped with screw-on terminals for 75-ohm cable, you can simply replace the old wire.

If, however, your set and antenna have terminals designed to handle only 300-ohm wire, matching transformers may be used to make the conversions, both at the antenna and at the televison set. The transformers have 300-ohm (twin-lead) and 75-ohm (coaxial) sides. The 300-ohm side is attached to the 300-ohm terminals of the antenna or television set, with the coaxial cable connected to the 75-ohm side.

You may need only one such transformer if either the antenna or the television set already has a 75-ohm terminal. Otherwise, you'll need two matching transformers—a 300-ohm-to-75-ohm and a 75-ohm-to-300-ohm.

The coaxial cable uses a screw-in or push-in connector fitting, shown in Figure 5-9, with the single center conductor of the cable as the male prong. This conductor sticks through a hole in the fitting, with the lattice shield attached to the outer (threaded) part of the connector. This outer metal portion is the *ground.*

Coaxial cable can be purchased in almost any length with the connectors already attached to each end. You can save money, however, by making your own cables. You can buy the cable by the foot and the connectors separately. No soldering is required if you get solderless connectors, which are simply crimped into place. (Although you can do the crimping with a pliers, an inexpensive crimping tool will do a better job.)

Note: Some television sets have both a 75-ohm *and* a 300-ohm antenna input. The 300-ohm input usually consists of two screw terminals for accepting spade lugs. The 75-ohm input will be a female RF connector mounted on the back of the television set. In rare instances, the television set will have a single 75-ohm female antenna connection with a slide switch beside it. The switch will be marked in "75-ohm/300-ohm" positions. This is an inefficient connector because you will need 75-ohm cable to feed it, and it will be a mismatch if a 300-ohm signal is being fed. Such a connector will work, but it will receive weaker signals when a 300-ohm twin lead is attached to the input, even if a matching transformer is used, because the connector itself offers close to a 75-ohm impedance even when switched to the 300-ohm position.

OUTSIDE ANTENNA CONNECTIONS

An outside antenna hookup can be connected to a television set in three basic ways: through use of a 300-ohm-to-75-ohm transformer (Figure 5-13A), a 300-ohm twin-lead antenna wire (Figure 5-13B), or

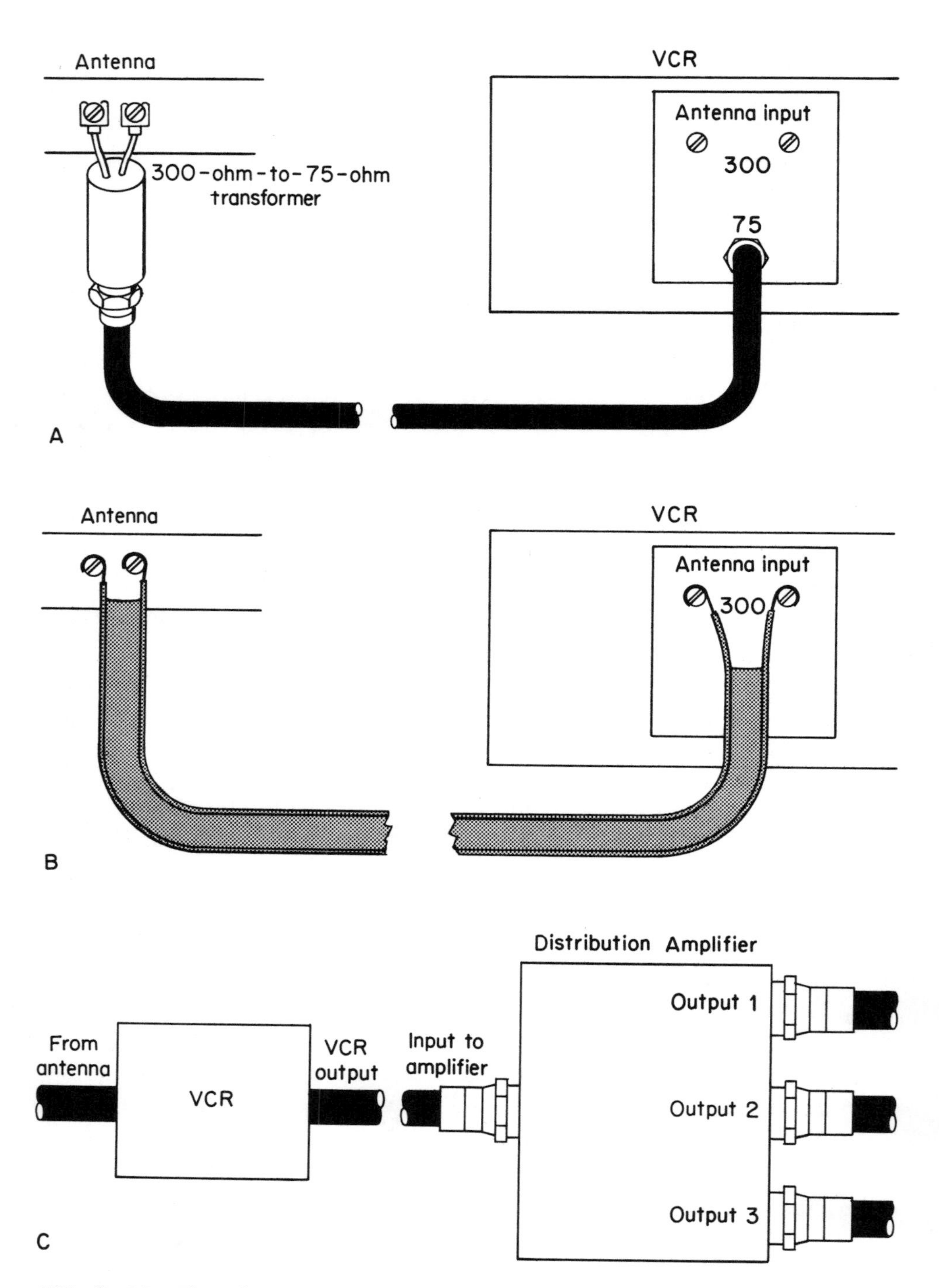

FIG. 5–13 Three basic connections to an outside antenna (*A*) 300-ohm-to-75-ohm tranformer (*B*) 300-ohm twin-lead antenna wire (*C*) Distribution amplifier.

a distribution amplifier (Figure 5-13C) (you may need a combination of these three, depending on your equipment). Look at the back of your set and compare it with the three installations shown in Figure 5-13. Make a note of your hookup for future reference in case you want to disconnect the VCR and reattach the antenna system only to your television set.

If your outside antenna is connected to your television set with a 75-ohm (round) coaxial cable, unscrew the RF connection from the back of the set. This connection will be either to an RF connector directly mounted on the set or to a 75-ohm-to-300-ohm matching transformer with spade lug-outputs that was provided by the antenna manufacturer. Take the 75-ohm cable from the antenna and screw it directly to the VHF input female connector on the back of the VCR.

Make the "VHF out to TV" connection exactly as shown for rabbit-ear connections. A 75-ohm RF cable goes from the "VHF Out"; the other end connects to the VHF input terminal on your television set.

How you complete the connection depends on your equipment. If you have the choice between 300-ohm and 75-ohm connectors on the equipment, making the connection is easy. A 300-ohm wire is attached to the screw tap labeled "300 ohm"; a 75-ohm cable is connected to the RF terminal labeled "75 ohm."

If your equipment doesn't offer you a choice, you will have to use adaptors to transform the impedance of the wires or cables to the impedance of the connectors on the equipment.

If your outside antenna is connected to your television set with a 300-ohm antenna wire and your VCR does not have 300-ohm input, connect the twin-lead spade-lug terminals to the screw taps of your 300-ohm-to-75-ohm VHF adaptor. Connect the other end of the antenna adaptor to the "VHF Input" on the VCR.

As before, connect the 75-ohm RF cable provided with your VCR from the VCR terminal marked "VHF Out" to the antenna input on the television set. If your television set doesn't have a 75-ohm RF input connector, you will need a 75-ohm-to-300-ohm matching transformer, which has the necessary spade-lug terminals to attach to the VHF input screws on the back of the television set.

Hooking up the distribution amplifier is much the same as hooking up a splitter or an RF switch. The input to the amplifier can be the signal coming from the antenna or cable, with the outputs connected to various television sets and VCRs. Or you can use the VCR to supply the input signal to the amplifier and feed a number of television sets through the amplifier output.

UHF CONNECTION

All of the hookups described thus far are only for VHF channels—channels 2 through 13. Many rabbit ears and outside antennas have provisions only for these VHF channels.

To incorporate UHF (channels 14 to 83), you will have to hook up a UHF antenna to your VCR and run a cable from the VCR's UHF output to the UHF antenna input of your television set.

Many VCR units have screw-type terminals on the back for 300-ohm twin-lead antenna-wire spade lugs, so matching transformers are not necessary. You simply connect a length of twin-lead antenna from the spade-lug UHF output terminals of the VCR to the spade-lug UHF input terminals of your television set.

When the VCR UHF input and output connectors are of the RF type, 75-ohm cable and the necessary matching transformers and antenna adaptors must be used in exactly the same way as in the hookup for VHF channels.

If your outside antenna uses a signal splitter (Figure 5–14) to provide both VHF and UHF inputs, then you already have the necessary four spade-lug outputs. A signal splitter is an inexpensive device similar to a miniature junction box, with one main input cable and two or more output cables.

Remove the spade lugs from the set and attach them to the VHF and UHF inputs of the VCR, using the appropriate connecting cables and matching transformers. Likewise, connect the VHF and UHF outputs of the VCR to the VHF and UHF inputs on the television set. These connectors allow you to videotape any program broadcast in your area, either on the VHF (channels 2 to 13) or UHF (channels 14 to 83).

FIG. 5–14 Signal splitter.

RABBIT EARS

An outside antenna, cable system, or satellite receiver ensures the best possible signal for your recordings. Using rabbit ears with a VCR is generally not a good idea. Rabbit-ear antennas must be repositioned each time you change the channel. If you have the rabbit ears set to record channel 10, the signal you receive on channel 12 may be too poor for viewing or recording. In addition, the quality of reception can change drastically whenever someone moves around the room, because the human body can affect television signals on their way to the rabbit ears.

If you must attach your VCR to a television with rabbit ears, disconnect the leads from the antenna input terminals on the back of the television set. You may have to extend these leads so that they can be attached to the back of the VCR. The output of a rabbit-ear antenna is 300 ohms, whereas the antenna input to the VCR is usually a 75-ohm coaxial connector. To make this connection, you will need a 300-ohm-to-75-ohm VHF antenna adaptor (transformer) with two screw terminals for affixing the spade-lug endings of the rabbit-ear antenna. On the opposite side, the adaptor will have either a male screw-on RF connector or a male slip-on RF sleeve.

To make the hookup, connect the spade lugs from the 300-ohm (flat) twin-lead wire coming out of the rabbit-ear antenna to the screw-in terminals of the adaptor. Then either push on or screw the other end of the adaptor into the VHF input connector on the back of the VCR.

Next, take the length of 75-ohm RF auxiliary cable that comes with the VCR. One end of this cable will go to the connector marked "VHF out to TV" on the back of the VCR. The other end attaches to the terminals that once accepted the rabbit-ear antenna. You will need a 75-ohm-to-300-ohm matching transformer to change the 75-ohm cable to the 300-ohm impedance of the television antenna. One end of the transformer is a *female* RF connector. Screw the end of the RF cable into the female connector. The other end of the transformer is a 2- to 3-inch section of 300-ohm (flat) twin-lead antenna wire, terminating in spade lugs. Slip these spade lugs under the terminals marked "VHF Antenna" on the back of your television set.

CABLE TELEVISION CONNECTIONS

The output of the cable system channel-selector box feeds to the antenna input connections on your television set. Disconnect these antenna input connections and attach them (with matching transformers if necessary) to the VHF and UHF inputs on the VCR. In turn, run cable and connectors from both the VHF and UHF outputs of the VCR to the UHF and VHF inputs on your television set.

If you are hooked up to a cable system where all station switching is done through the cable company's channel-selector box, then set the VCR to the same channel as your television set.

A more complex setup involves switching between the VCR tuner section and the cable-selector box. This is possible *only* if the VCR has a "cable ready" tuner and if the "unscrambling" has been done before the signal reaches the VCR. To do this you will need a special switch designed to handle the high frequencies involved (Figure 5-15).

A two-way signal splitter receives the incoming signal. One output goes to the selector box; the other goes to the VCR. The outputs of the selector box are then cabled to the switch, with the output of that switch connected to the television set or the VCR (Figure 5-16A).

You can now view one channel while recording another. With slight modifications, this wiring scheme can be used to do the same thing with either antenna or satellite receiver input.

You can use an RF switch to control a remote television set in another way. Two cables are run from the main location to a switch at the remote television set. One cable carries the cable television signal; the other carries the VCR signal. These cables are connected to the same side of the switch. The signal connector on the switch is con-

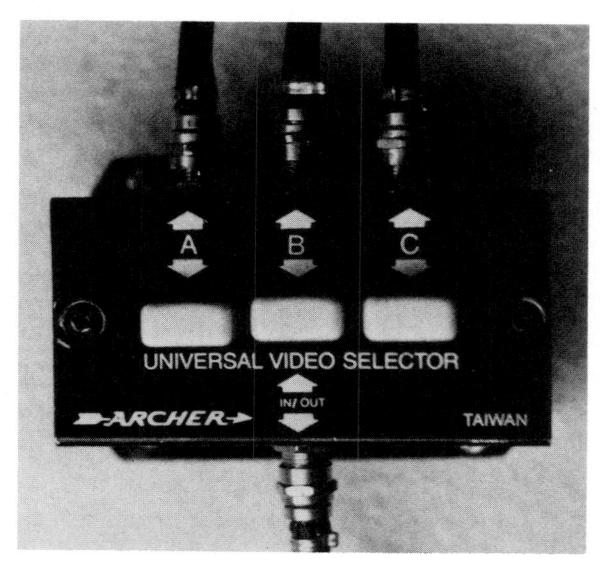

FIG. 5–15 RF switch.

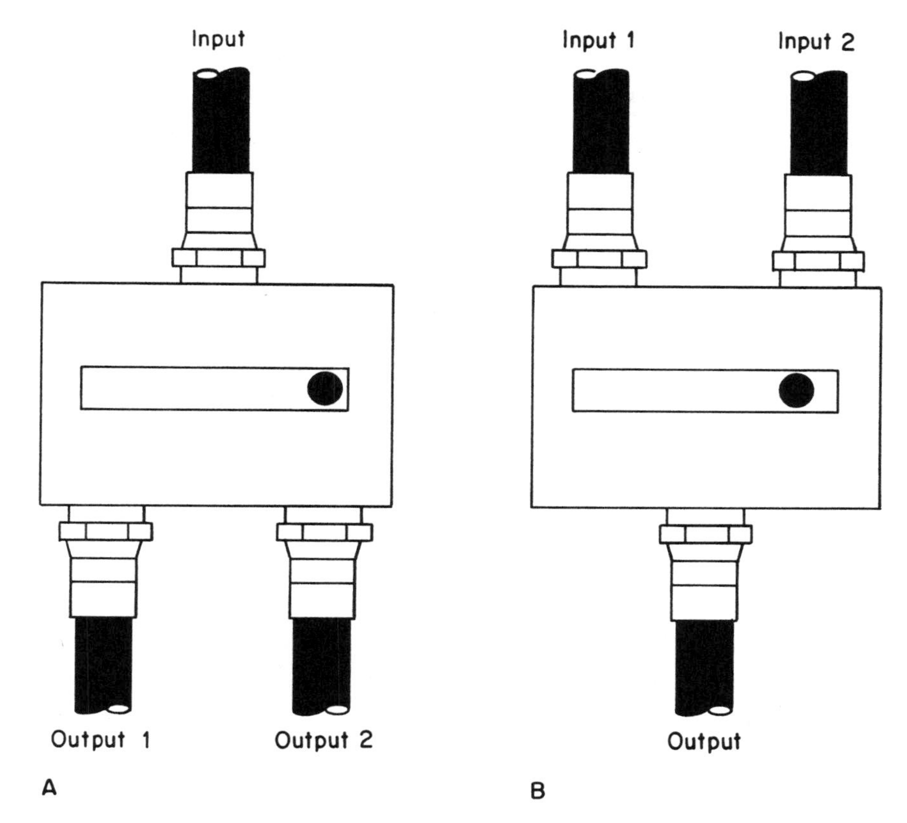

FIG. 5–16 Hooking up the RF switch.

nected to the remote television set. You now have a means of switching the remote television set from regular cable television input to VCR input (Figure 5–16B).

MORE THAN ONE TELEVISION SET

The easiest way to drive more than one television set is to use a signal splitter. Signal splitters can be purchased with either 75-ohm or 300-ohm terminals or a combination of the two, usually with built-in matching transformers. The signal splitter is usually installed at the first television set. The main line comes from the antenna or cable system and is attached to the input of the splitter. Then cables are run from the outputs of the splitter to the individual television sets. For maximum efficiency, use 75-ohm coaxial cable.

Another method, which is more expensive but much more effective, is to use a "distribution amplifier" instead of a splitter. The signal is then both split and boosted. If you have more than three television

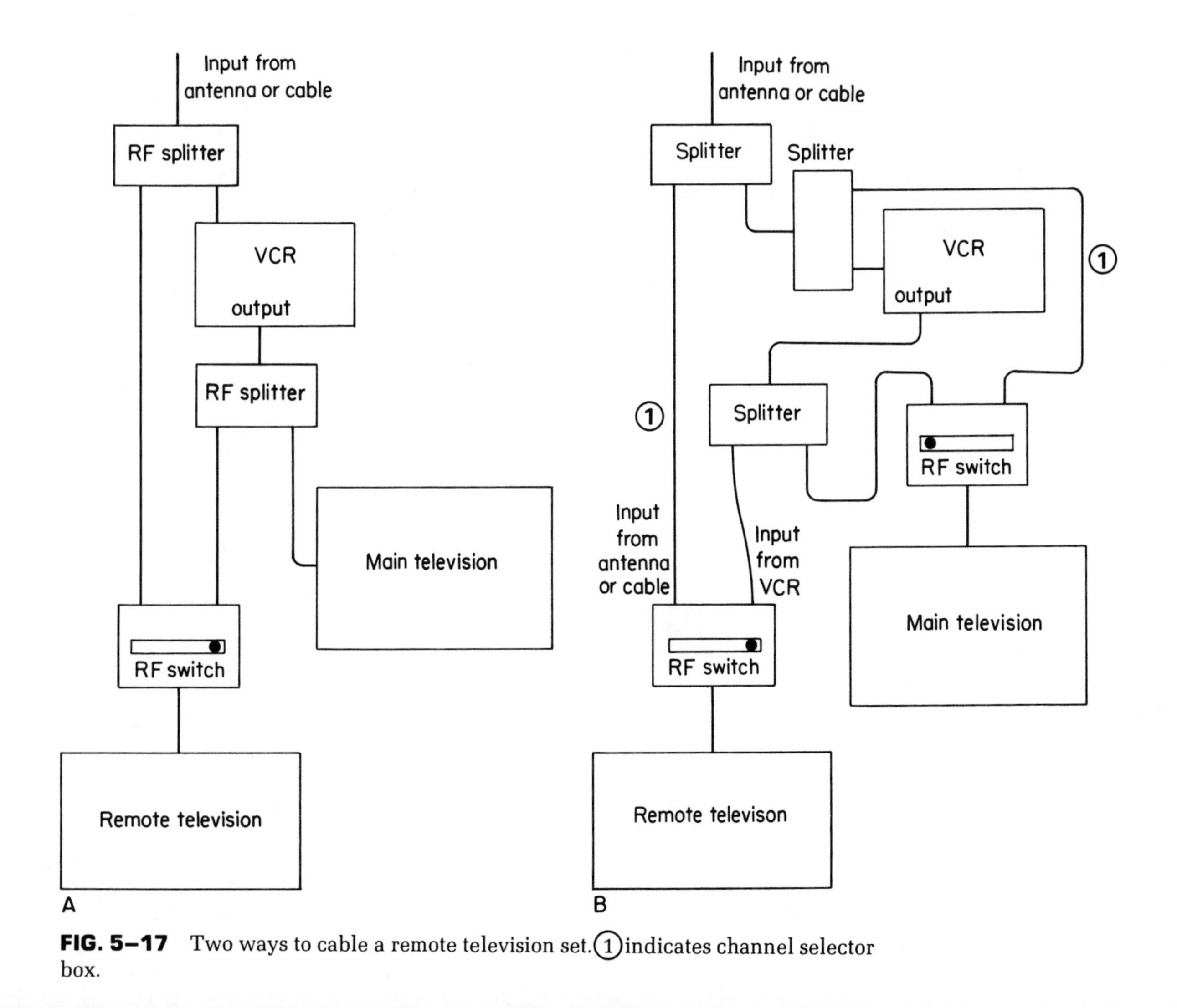

FIG. 5–17 Two ways to cable a remote television set. ① indicates channel selector box.

sets on a single antenna system, you may need a distribution amp. You can buy one at an electronics supply store for between thirteen and several hundred dollars, depending on the power, complexity, and quality. The distribution amp must be plugged into an electrical outlet. The amplifier draws very little current and can be plugged into almost any circuit without fear of overloading breakers or fuses.

Hooking up a distribution amp is simple. The antenna lead-in cable is connected to the input of the amplifier. Then 75-ohm coaxial cable is run from the outputs of the amplifier to each television set.

If you want to have control of the remote set—to be able to watch television while someone in the main room is watching a videotape—you will need two cables (see Figure 5-17). One cable carries the television signal; the other carries the signal from the VCR. An RF switch is needed at the remote site. The two cables are attached to the two input terminals, with the output terminal connected to the remote television through the switch.

You can now select which you want to watch. Flip the switch to side "A" for the normal television signal or to side "B" for the VCR.

SUMMARY

Most VCR connections employ a spade lug, an RCA pin jack, or an RF connector. You can buy the wires and cables or save money by making them yourself. To save time and trouble, use solderless connectors.

Most of the connections between television sets and VCRs, or between VCRs and VCRs, require little more than common sense and planning. By using signal splitters, switches, and some ingenuity, you can do anything you wish. Distribution amplifiers can be used both to split and to boost the signal if you have multiple television sets.

The best wiring is done by using cable. The 300-ohm twin-lead antenna wire can lead to degradation or loss of signal. Cable is much more efficient and makes for a neater installation.

Keep things simple. When you've sketched the plan for what you have in mind, see if you can simplify it. Often you'll find that your original idea does things the hard way.

Make a complete sketch of your connecting scheme. Sooner or later you'll be glad that you have it as a reference.

Chapter 6

The Tapes

The introduction of cassette tapes to the video industry proved a major boon, since it allowed manufacturers to mass-produce machines that are extremely simple to operate. In fact, modern VCR units are so nontechnical in operational design that using one is as simple as dropping a letter in a mailbox.

VCRs are either top loading or front loading, but the cassette is the same for both.

There are some slight differences, however, between Beta and VHS, most notably in the size of the cassette.

Not only is the video cassette easy to use, but it also helps protect the tape. When the cassette is pushed into the slot, a mechanism unhinges the front flap, pulls the tape out, and winds it along the tape guides and across the heads. This is done automatically; you never have to touch the tape.

Many VCR owners have been forewarned that the most frequent causes of malfunction are dirty heads or careless operation that has allowed dust or moisture to accumulate inside the machine. But most users are not aware that more than half of VCR problems are actually caused by faulty cassettes.

The cassette is a delicate mechanism. To keep the price low, manufacturers use a lightweight plastic that is not very sturdy. The spring-loaded protective flap, or the plastic hubs or ratchets that allow the tape to be moved back and forth inside the cassette, can easily become broken or shift out of alignment. If the cassette is defective, the machine will not be able to load the tape. A bad cassette can even damage the heads and the machine.

Precautions

Avoid moisture.
Don't use a cassette until it has reached room temperature.
Keep the cassettes in their boxes and away from dust.
Store cassettes vertically.
Don't bump or jar the cassettes.
Keep cassettes away from magnetic objects.
Store in a cool, dry place.
Rewind completely before storage.
Don't touch the tape.
Fast-forward and rewind all tapes that have been in storage for long periods.

Most of these commonsense precautions concern video cassette storage. Improper storage can ruin the tapes and even damage the VCR. Proper storage will extend the life of your tapes and machine.

If you rent a movie on a cold winter evening and immediately bring it into a warm house and play it in a warm VCR, you're asking for trouble. The result is the formation of dew or condensation, which can lead to immediate deterioration of the cassette and the VCR. To keep moisture at a minimum, keep the video cassette at room temperature or allow it to warm to room temperature for at least an hour before using it. This is especially important when the cassette has been exposed to cold. In addition, moisture may be transferred from the tape to the delicate recording/playback heads and other inner mechanisms of the VCR.

Always store cassettes in the cardboard or plastic boxes they came in. For extra insurance, buy a video cassette storage case, preferably one that has doors or drawers that close (see Figure 6–1).

The cassettes should be stored vertically (on edge), never horizontally. Naturally, take care to prevent shock damage to the cassette. Don't drop them or bang them around.

Keep the cassette away from any magnetic field. We know of one person who set a cassette containing a favorite movie on the speaker of his stereo. When he went to play the movie, he couldn't figure out why it was suddenly erased. Electric motors, power transformers, and

FIG. 6–1 Tape storage case.

other devices generate magnetic fields. Keep the cassettes well away from suspected sources of magnetism.

Store cassettes *only* in a cool, dry place. Cassettes are sensitive to moisture *and* to heat. More than one owner has lost a valued recording by leaving the cassette in the car on a hot day. Others have damaged their prized recordings by carelessly placing them on radiators or too near heater vents. Some have found themselves having to pay for rented movies accidentally destroyed by such carelessness. Even if the heat doesn't melt the cassette, it can harm or ruin the recording.

Fully rewind a tape before storing it. DO NOT store tapes for more than short periods unless you have rewound them.

If a tape has been stored for a long time, or if you rent a tape that may have been used on an improperly maintained machine, fast-forward the cassette all the way to the end. Then rewind it before using it yourself. This is also important for new cassettes. Fast-forwarding and rewinding repacks the tape properly and helps avoid any sticking or moisture problems.

A good investment is a separate rewinder that does nothing but fast-forward and rewind. This preserves the motors of your VCR. Even a unit that does nothing but rewind can reduce the amount of wear on the VCR motors.

Separate rewinders can be as complicated as you wish. They can be battery-powered or plug into the wall. (AC-powered units are preferable.) Some just rewind, while others can go in either direction. Some have built-in erase heads. These work much better than the typical erase heads in a VCR. Some manufacturers claim that the erasing returns the tape to almost original condition, leaving it signal-free.

A new kind of rewinder now available safely cleans the tape each time you move it through the device. This idea has been around for some time, but previous units usually cleaned just one side of the tape, which is rather useless. The clean side will immediately come into contact with the unclean side—and so much for the cleaning.

NEVER touch the surface of the tape. This is one reason for the hinged flap on the cassette. Fingerprints and oils from your skin can ruin a tape very quickly. Dirt and grease can also damage the delicate internal mechanisms of the VCR.

ANATOMY OF A CASSETTE

The cassette or cartridge is a protective plastic box housing the tape and its two reels (Figures 6-2 and 6-3). The tape is protected by a hinged flap at the back edge of the cassette. When the cassette is not in the machine, a spring-load keeps this flap down, protecting the tape from fingers and other dangers. On one side of the cassette is the release catch for the hinged flap. The VCR automatically presses this button to release the flap and allow access to the tape. You can also release the flat manually by pressing this button.

FIG. 6–2 VHS and Beta cassettes. The larger one is the VHS format.

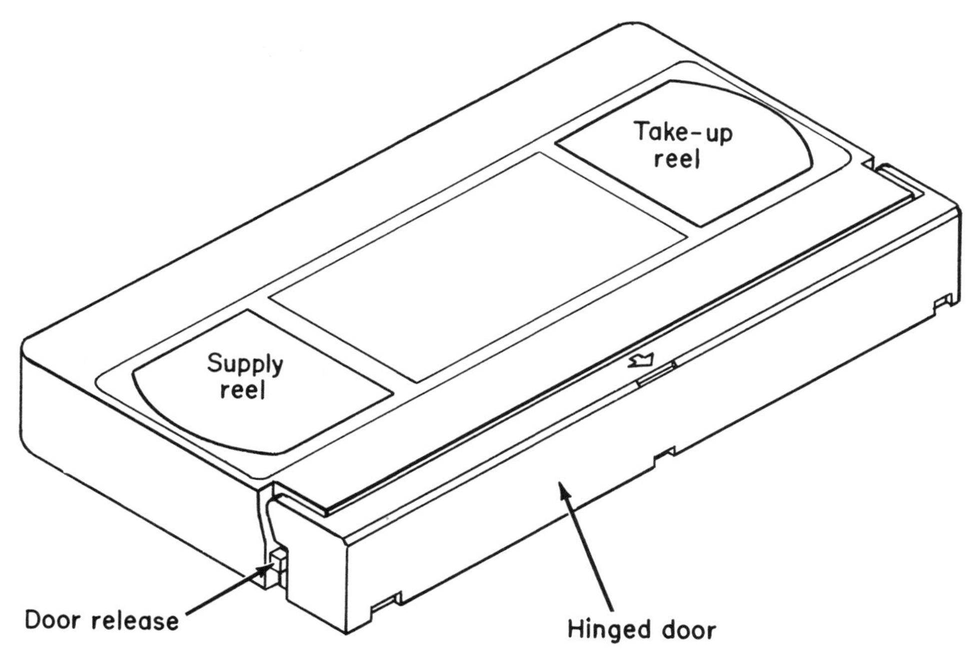

FIG. 6–3 Top view of a cassette.

The underside of the cassette (Figure 6-4) has a small hole that allows a pin from the machine to slip into the cassette. The door is released, and the tape is fed into the VCR.

Warning

Cassettes must be inserted into the VCR with the label up. They should not be inserted upside down.

To prevent accidental erasure of a recorded program, all cassettes have a removable tab on the rear edge opposite the hinged door. This tab can be broken off by pushing or prying it with your finger or with a screwdriver. Very little force is needed to break the tab. After the tab has been removed, the cassette can be used only for playback, and any material that has been recorded is protected from being erased. If you want to record on the cassette, merely cover the hole with a piece of tape.

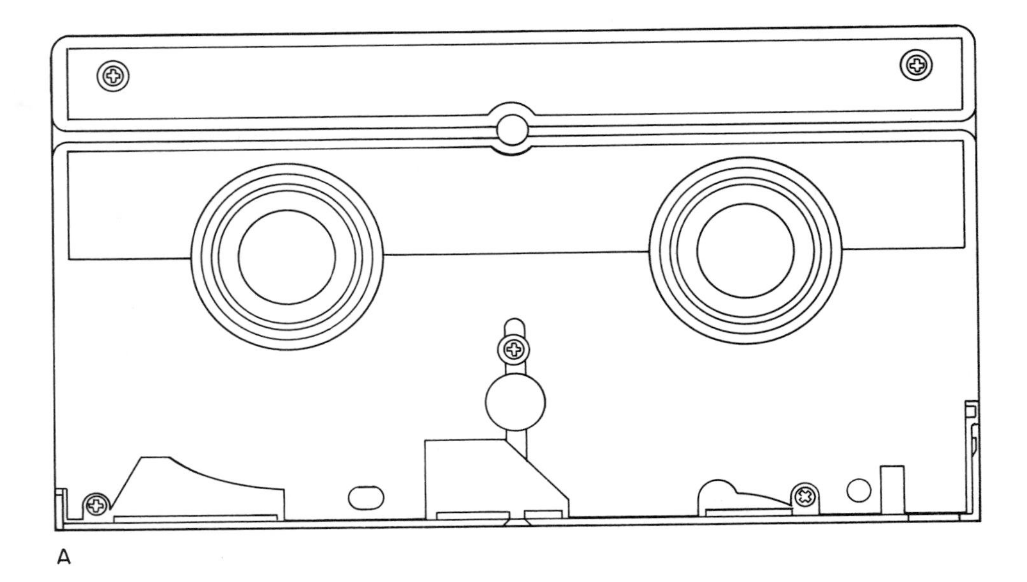

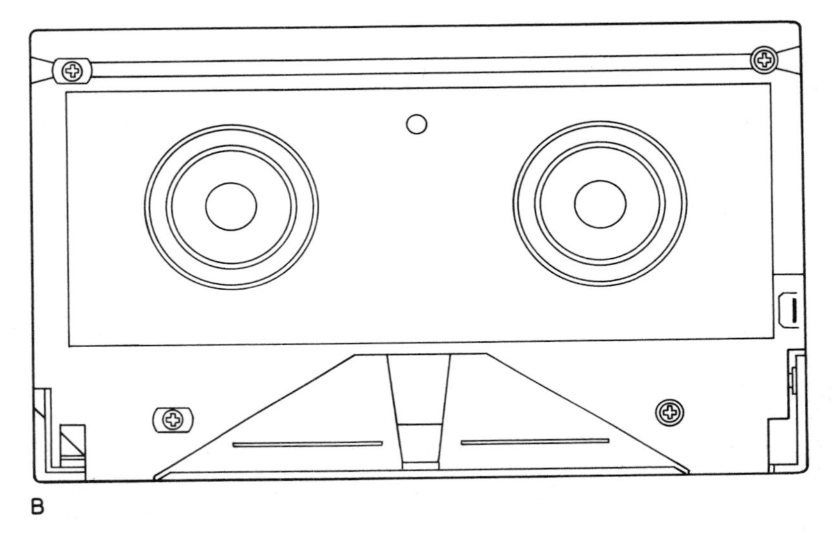

FIG. 6–4 Bottom view of a cassette.

TAPE REPAIR

Virtually all operating manuals, instructions that come with VCRs, and books on VCRs warn that cassette tape cannot be spliced as audiotape can be. Operators are warned not to attempt to repair the cassettes or tapes, or even to open the cassette. This is a fallacy. If a tape breaks near the end and it contains some valuable recorded material, much of that recorded program can be saved. Even if the break is in the middle, you can save much of the original recording. And you may be able to save almost all of it.

Most cassettes are held together with six or seven small Phillips-head screws driven through the bottom of the case. If these screws are removed very carefully, you can take the tape case apart. Make note of any springs or plastic tabs that are held in place by the pressure of the lid against the base (Figure 6–5).

Unwind the tape from the nearly empty spool and throw it away (Figure 6–6). Notice that it is attached to the plastic hub. Use a sharp scissors and trim the end of the tape to be saved to a square end (Figure 6–7). Use a piece of adhesive or electrician's tape to attach this free end to the empty reel (Figure 6–8). Next, carefully reassemble the cassette, being sure that all springs and tabs are back in their proper positions before putting the two halves together (Figure 6–9).

Before trying to play back or record a repaired cassette, be sure to fast-forward and rewind it from end to end at least *twice*. This will help the tape to resettle and repack.

Don't throw away a cassette just because the tape itself may be damaged beyond repair. The tape may be bad, but the cassette case may be in good condition. Rather than throw away the bad tape, you can wind it inside another cassette case and discard that tape instead.

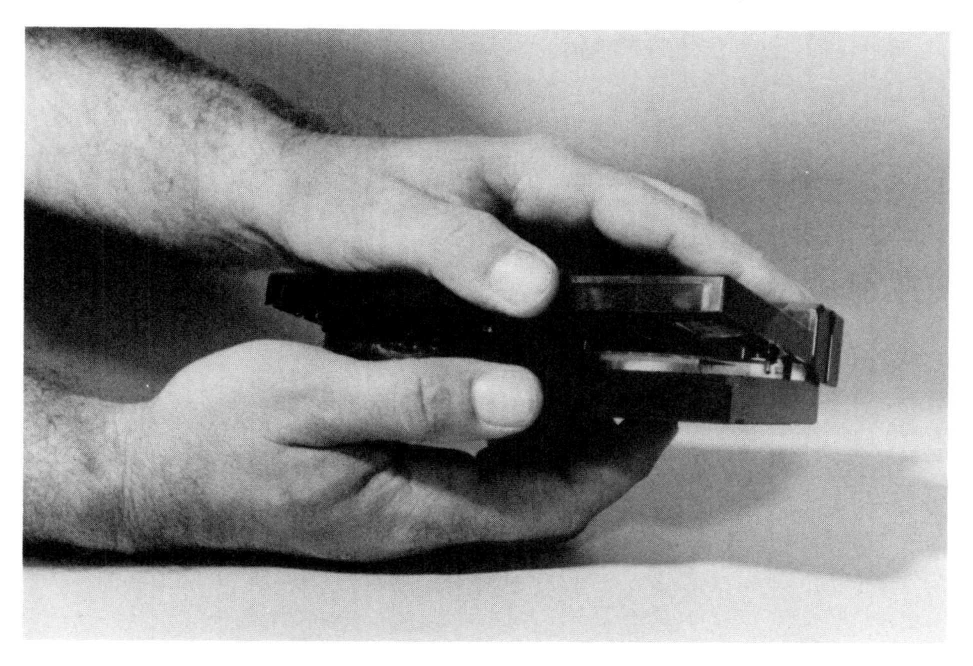

FIG. 6–5 Separate the base from the lid, taking note of the location of any springs or tabs.

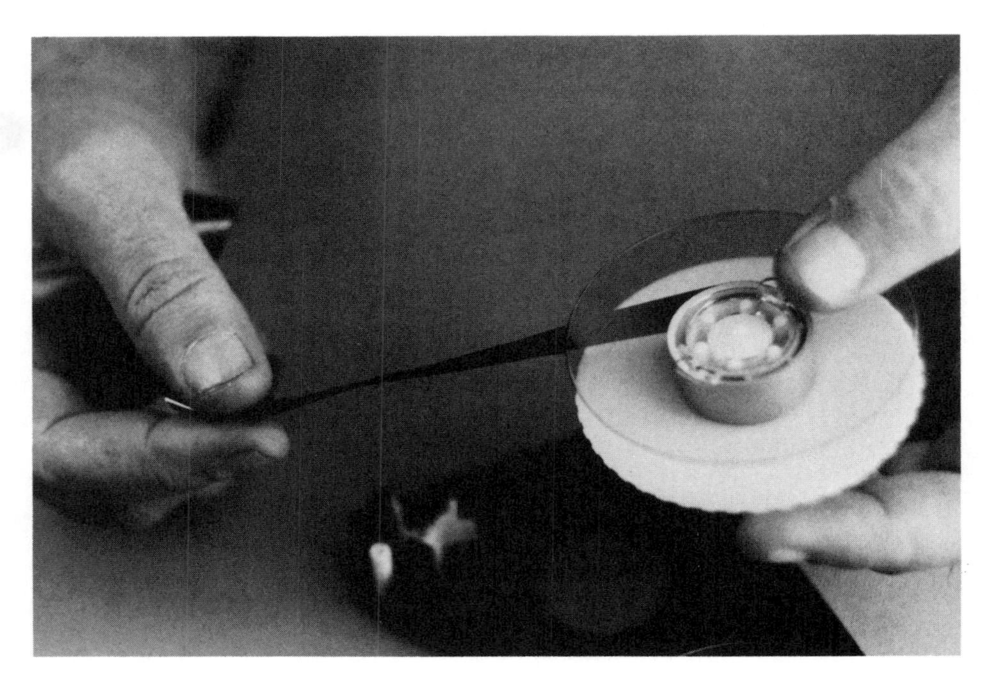

FIG. 6–6 Remove the tape from the emptier spool.

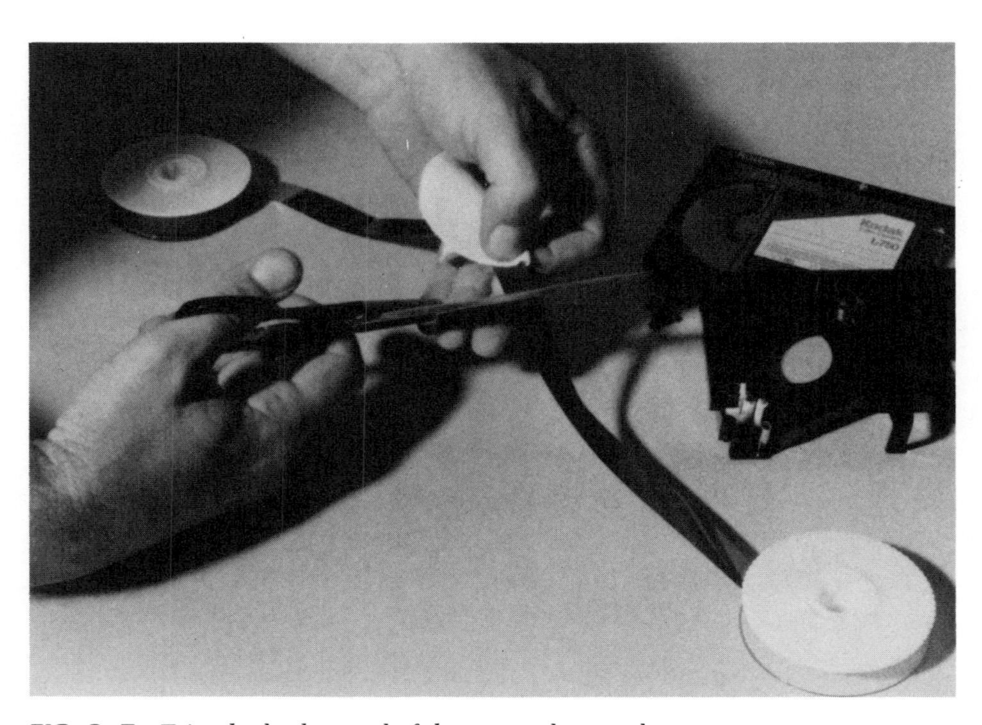

FIG. 6–7 Trim the broken end of the tape to be saved.

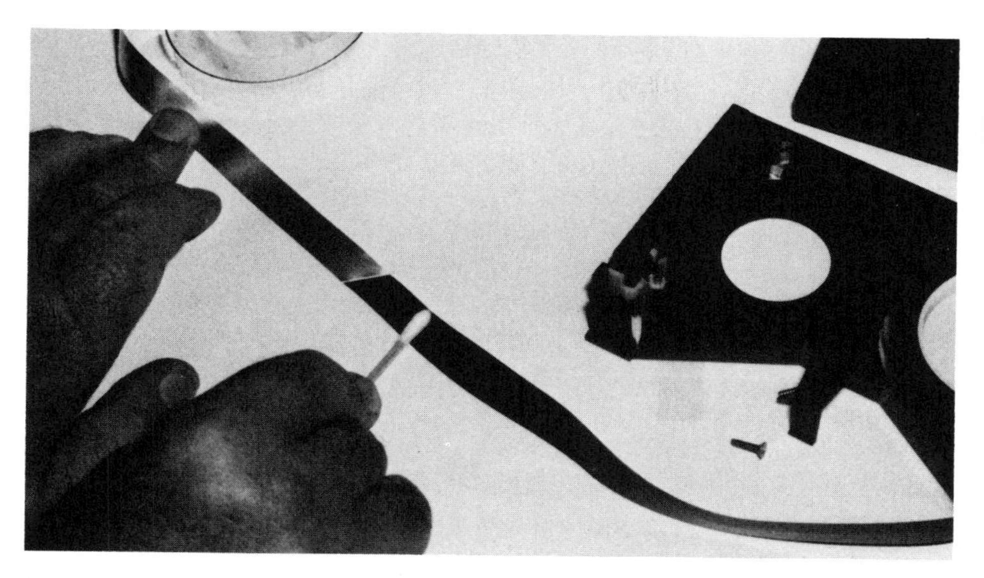

FIG. 6–8 Attach the tape to the leader.

In many instances, the springs, tabs, or support posts inside the cassette may have become bent, broken, or knocked out of position. In this case, you can save the tape by placing it in another cassette case. To save the tape, carefully disassemble the bad cassette, remove the two spools, and insert them into the good cassette case. (If the recording is important and you don't have an empty cassette, you may even want to buy a new cassette and swap the tapes.)

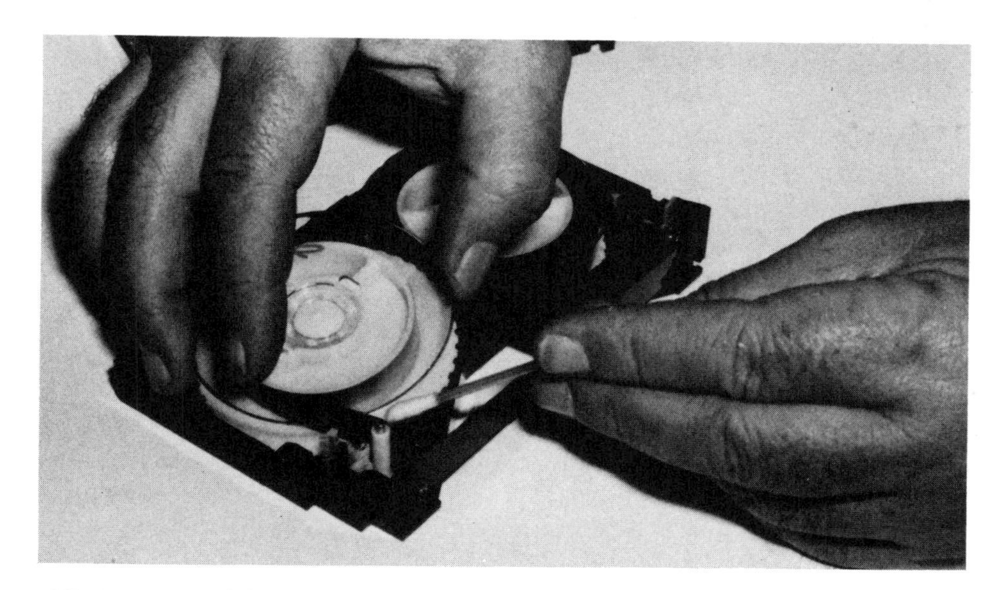

FIG. 6–9 Carefully reassemble the cassette.

SPLICING TAPE

Despite the warnings about splicing, a broken or torn videotape may be cut on an angle and spliced with audio splicing tape (in the same manner as audiotape). But keep in mind that you could face some serious consequences.

Whenever the splice crosses the record/playback heads, there will be a momentary signal dropout. You will lose part of the material you are trying to record or play back, but this is better than losing the entire tape.

More important than the dropout is the way the splice affects your machine. Every time the splice passes over the head, there is a chance that the cut tape edge may catch and damage the record/playback heads. The only time you should attempt to splice a videotape is if the material on the tape is more important to you than the VCR itself or the possible repair bill to replace the heads. Even then, make a copy of the tape right away to avoid future damage. (Each time the spliced tape is played, the chance of damage increases.)

Warning

Although splicing is possible, try to avoid it. It should be only a last resort. Splicing can damage the VCR heads.

Splicing requires great care. Wear a pair of lint-free gloves so that you won't actually be touching the tape. Or at least wash your hands well with soap and water.

Make the two cuts and be sure that the two ends are perfectly clean and flat. Then cross the two ends over each other and make a straight diagonal cut so that the ends mesh perfectly. Line up the two ends so that they are perfectly flat against each other and the tape is in a perfectly straight line. Finally, attach the two ends with splicing tape.

THE CASSETTE-LOADING MECHANISM

When a cassette is inserted into a machine, whether front loading or top loading, the tape is pulled out of the cassette and threaded in the machine through an arrangement of levers. The entire operation is completely automatic. Because of this automatic feature and the simplicity of the cassettes, the mechanics of the VCR unit are more complicated than those of reel-to-reel videotape machines. A complex system of levers, rods, and servomechanisms makes up the mechanics of the VCR. Adding to this complexity, the number of electronic circuits that control the mechanics makes the VCR more difficult to service than the older, more basic units.

When a cassette is loaded, the protective door of the cassette lifts open and the cassette is lowered over a vertical metal pin. This pin is on an arm or lever attached to a large circular ring called a *threading ring*. When the VCR is used for playback or is in fast-forward, the ring rotates and allows the arm and pin to thread the tape over the heads. When the threading has been completed, the threading motor stops and a solenoid clamps the pinch roller against the capstan. The tape now can move in its normal direction along the tape guides and across the rotating head.

While the cassette is threading, don't touch the other controls. In a sense, this would be like trying to force the tape and mechanisms in two directions, or two speeds, at the same time. Most VCRs have protective devices that prevent the controls from operating while the tape is threading, but it is still possible to damage the machine or tape by pushing the selector switches at the same time. Don't trust the protective devices.

Once the threading is complete, the function buttons operate and there is no danger of damaging the tape or machine. Threading usually takes only a few seconds.

Whenever you push a stop button, the threading ring rotates, unwinding the tape from across the heads and the tape guides, then places it back inside the cassette. In some models, the machine can then be fast-forwarded or fast-rewound without the tape crossing the heads. In newer models with stop-frame and other specialized functions, the tape might stay in its ready position around the rotating heads regardless of the mode of operation. Because of this, video cassettes should be removed from the VCR when not in use. When the stop button is hit at the end of a playback or record function, the cassette should be removed before the machine is turned off.

UNTANGLING TAPE

If a circuit becomes defective or the controls malfunction, the stop button may *not* cause the machine to halt and the tape to go back into the cassette. Instead, it may cause the tape to become unthreaded and pulled out of the cassette again, and as a result, previously recorded material may be recorded over. This can go on and on, ruining the end of a program you've taped and also possibly damaging or destroying the end of the tape. If this occurs, turn off the power or unplug the VCR. The longer the machine continues to run, the worse the tangle will be.

A tangle inside a top-loading machine is much easier to fix than one in a front-loading machine. With a top-loading machine, you can often get at the tangle merely by removing the small top cover. With a front-loading machine, and with some top-loaders, you may have to remove the top part of the cabinet completely to get at the tangle.

The protective dust cover of a top-loading machine is usually held by about a half-dozen Phillips-head screws or normal slot-type screws. If you remove these screws carefully, noting which goes into which slot if they are of varied length, you can carefully lift off the top.

If a malfunction occurred because of an improperly functioning circuit or switch or a damaged piece of tape, carefully untangle the tape until you find a loose end, free of all the machinery. (Whenever you must handle tape, wear clean, lint-free gloves.)

Plug in the machine, put it in stop mode, and turn the power on. The unthreading process will tend to pull the tape back into the cassette. Guide the tape to keep it from becoming entangled again and from touching the rotating heads. When all of the motors have stopped, try to eject the cassette.

Once you have removed the cassette, check it carefully for any crinkling or other damage. The bad tape should be pulled out from both reels through the hinged flap and cut off.

Cut off the portion of tape that you have touched with your fingers, even if it is undamaged and the machine is able to draw it into the cassette. Grease or oils from your fingers can damage the rest of the tape or the delicate heads of the VCR.

Now you have two choices: You can either throw away the shortest side of the tape and reattach the good end from the rest of the tape to the other reel, or you can try to splice the tape.

BUYING TAPES

The best grades of tape are made for video and delicate computer recording media. Sometimes lesser or economy grades will work in a

VCR recorder but will not last as long as better-quality tapes, and they are more likely to cause excessive wear to the heads and transport mechanisms.

Using high-quality tape will lengthen the life of your machine. Better tape will also last longer and cause fewer problems. Stay away from so-called bargain brands.

According to a recent article in *Consumer Reports*, just because a tape *says* that it is of higher grade doesn't necessarily make it so. The highest marks were given to standard-grade Scotch videotape. It even outperformed the higher-grade tape from the same company. (This testing was done only on VHS machines.) Other brands rated as being excellent were Fuji and TDK. The only brand-name tape that didn't seem to perform well was PDMagnetics. (This testing was done in mid-1984. Formulations, manufacture processes, and even the manufacturer may have changed since then.)

So, stick with the known brand names for reliable tapes and cassettes. If you buy an unknown brand, you're taking the chance on inferior quality.

SUMMARY

Tape quality can make the difference between reliable, enjoyable recording and playback and headaches. Of all the problems with VCRs, more than half are caused directly or indirectly by the cassettes.

You can greatly reduce the problems and extend the life of your tapes by taking care of them properly. Handle them carefully at all times. Never touch the tape with your fingers. Most important, store them correctly.

If a tape has to be repaired, you have several choices. If the cassette itself has gone bad, the easiest repair is by simply swapping the tape from the bad cassette case to a good case. For more serious problems, the tape can either be remounted to the reel, with the bad part tossed out, or be spliced. Splicing videotape is a last resort, to be used *only* for irreplaceable recordings. The splices can cause serious damage to the record/playback heads of the VCR.

Tangled tape can be worked out of the machine. Sometimes the tape will not have been damaged by the tangle.

Stay with the known brands of cassettes. The bargain brands are usually no bargain at all. At best, you'll get a poor-quality recording and a short tape life. You can also cause expensive damage to the machine.

Chapter 7

Periodic Maintenance

Since cleanliness is the key to proper maintenance, the most important technical knowledge you can gain is to learn the proper steps for thorough cleaning. Once you master these steps, you can greatly reduce the frequency of malfunctions.

Most of the steps for proper maintenance are nothing more than common sense. Yet the lack of routine maintenance is the major reason why VCRs malfunction or stop working entirely.

If you ignore our constant badgering to "Keep it clean," you'll find yourself buying a new machine much sooner than you'd expected.

KEEPING THE VCR CLEAN

Daily removal of dust and moisture is a must. Simply use a barely damp dustcloth or one impregnated with special dust-lifting chemicals to wipe exposed surfaces. Follow a damp cloth with a clean, dry rag. This will help keep the dust from filtering into the machine. Don't use a cloth to clean the inside of the machine. Buy or make a dust cover and cover the VCR when you aren't using it. Don't forget that a surprising amount of dust can gather in just a few hours, and it has the amazing ability of getting in exactly the wrong places. Given the chance, it will seem to zero in on the most delicate parts of your machine.

The best way to avoid moisture is to keep the machine at a constant temperature. Condensation can occur whenever the temperature in the room vacillates between warm and cool. Condensation on the outside of the cabinet is not a problem. If it's inside, however, even invisible droplets of moisture can cause serious damage.

Warning

Do not operate the machine with the dust cover in place. This will block ventilation and cause overheating that will lead to equipment failure.

Some VCRs have a device called a dew sensor. This device will shut down the machine if the humidity level rises high enough to be potentially harmful.

If you live in an area where the humidity is high, packets of silica gel can help absorb the moisture before it does damage. These packets are inexpensive and reusable. (After they have absorbed their quota of moisture, bake them in an oven at low heat for a short time to restore them.)

The packets are usually made of paper or fiber, neither of which will conduct electricity. You can place them anywhere inside or outside the unit, but be careful to keep them well away from any moving parts (see Figure 7-1). Inside a VCR this isn't as easy as it may sound.

FIG. 7–1 Be sure that the silica-gel packet is secured in place away from all moving parts.

Take your time and be sure that the packet is in a safe place. Tape the packets in place so they won't shift around if the machine is moved.

A large ore freighter was moving through the Great Lakes. As it approached one of the drawbridges, something unexpected happened. The ship kept coming, and the bridge refused to lift. At the last possible moment the operator was able to get the bridge to move by a manually operated control. This near disaster was caused by a moth killed by the contacts that operated the bridge-raising circuits.

While dirt and foreign objects will not cause a disaster of this scale for your VCR, they can cause serious mechanical and electrical problems. Each time you open the unit, check the interior for cleanliness. Not only can dust build up, but insects can fall or crawl inside.

If your machine has plug-in circuit boards (Figure 7-2), the contacts may acquire a coating of either grime or corrosion. This doesn't happen often, but when it does it can cause serious malfunctions or erratic behavior. Often, merely unplugging and reinserting the circuit boards will solve the problem. (Do this with the POWER OFF.)

If this doesn't work, clean the contacts at least once a month. If you use a good cleaning fluid, you won't wear off any appreciable amount of the metal on the contacts. The cleaning fluid can be any

FIG. 7–2 A plug-in circuit board.

FIG. 7–3 Cleaning the contacts.

contact cleaner, freon, technical-grade isopropyl alcohol, or even the soft eraser of a pencil (see Figures 7-3 and 7-4).

There are three main precautions. First, the cleaning fluid should contain no lubricants. Second, if you must resort to a pencil eraser, clean the board well away from the machine and thoroughly wipe off the contacts. Rubber eraser shavings can cause more problems than you had in the first place if they get in the wrong places. Third, always wear lint-free gloves whenever you must touch any interior parts. You can get these at many electronics supply stores or even at a local camera store. (Photographers use them for handling negatives.)

CLEANING THE HEADS

As the tape goes through the VCR and across the heads, all sorts of things happen. The tape can leave tiny amounts of dirt, oxide, and binder behind as it is being played. As these build up, the gap between the tape and the heads increases, causing a loss in image quality. Let this buildup continue and it can become a permanent part of the heads. The only thing you can do if this happens is replace the heads.

FIG. 7–4 An emergency procedure: You can use a pencil eraser if nothing else is around.

As usual, this doesn't have to happen. All you have to do to prevent such a costly repair is to keep the heads clean. The first step is to keep the machine and surroundings clean. Second, be sure to use only high-quality tapes. Third, make head cleaning a regular part of your maintenance routine. The heads should be cleaned after each 25 to 50 hours of operation—for example, after each 10 movies. Unless the VCR has not been used and has been kept covered, plan on cleaning the heads once a month.

The easiest way to clean the heads is to use a special head-cleaning cassette. However, these are inferior to manual cleaning and not as efficient as a cleaning pad and fluid, since they don't clean the tape guides or other transport mechanisms effectively.

Head-cleaning cassettes have also been known to leave behind harmful deposits. One of the "bargain brand" kits can actually make the machine dirtier. Even worse are the ones that attempt to scrub the heads. This abrasive action severely shortens the head life.

Although head-cleaning cassettes are not recommended over manual cleaning, they are still better than nothing. If you're uninclined to do the job by hand, use the highest-quality cleaning kit you can find. But expect to have problems sooner or later.

Some VCR owners use a cassette for quick, occasional cleaning, then every few months they take the time to open the cabinet and do the job thoroughly by hand.

Cleaning the heads manually is not difficult. It means that you'll have to remove the top cover to gain access, but this takes just a few minutes. You'll need cleaning fluid (alcohol or freon, for example, but NEVER one that contains lubricant) and cleaning pads. Do not use swabs on the heads because they can leave behind tiny fabric hairs.

Before attempting any cleaning, shut off the power. Some people are tempted to leave the power on, with the idea that the rotating-head drum will clean itself as the pad is held stationary. This is a great way to destroy the heads.

Exert only minimal pressure when cleaning the heads. The cleaning pad should just barely touch the heads. The motion of the pad against the head is always horizontal (around the head), never up and down. Exerting too much pressure or using an up-and-down motion is almost certain to damage the heads or knock them out of alignment.

The simplest method for cleaning the heads is to hold the fluid-impregnated pad *lightly* against the head drum with one finger while turning the drum by hand from the top (Figure 7–5). A second method is to move the cleaning pad around the drum while holding the drum in place with a finger. Either way, be sure to wear a lint-free glove on the hand that touches the drum.

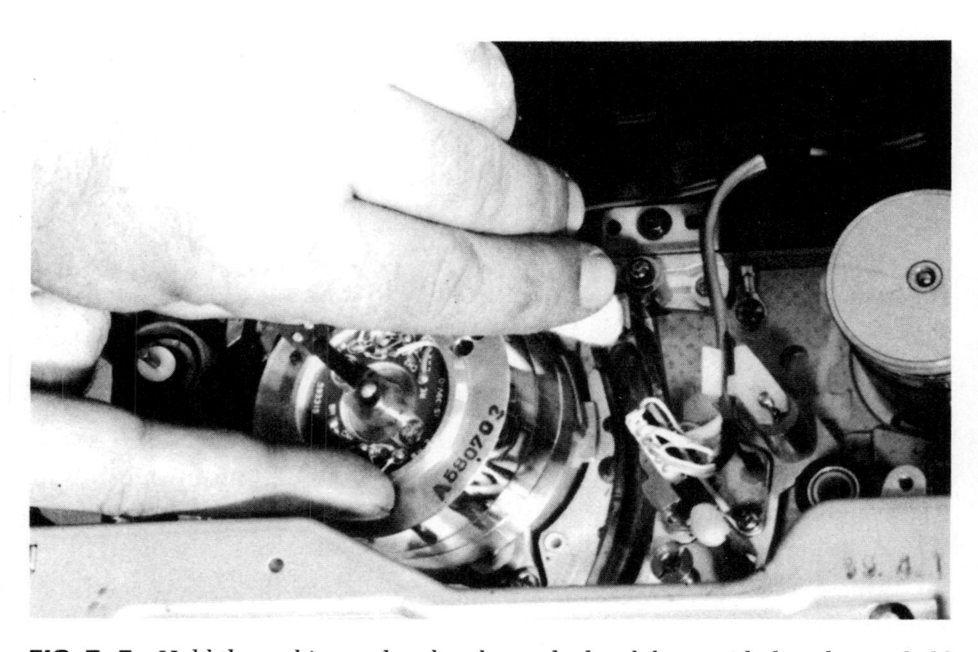

FIG. 7–5 Hold the pad in one hand and turn the head drum with the other, or hold the drum and move the pad around the head drum.

If you clean the heads after every 25 to 50 hours of operation, the dirt and grime will come off easily. However, given too much time between cleanings, those residues can bind themselves to the head, possibly permanently.

CLEANING OTHER PARTS

As the tape is pulled from the cassette, it is fed through a series of six to twelve tape guides. These are usually upright tracking pins made of metal or plastic. Closely associated with the tape guides are the various rollers and capstans and the threading ring. (See Chapter 6 for a detailed discussion of the way tape moves through the VCR.)

All these parts can also become coated with deposits of oxides. In turn, the tape itself can pick up the contaminants, making the problem of a dirty mechanical tape transport a vicious circle.

Clean all parts and places where the tape touches, and do this regularly (Figure 7-6). Clean everything each time you clean the heads. You can use cotton swabs on these parts if you wish, but cleaning pads are better because the swabs can leave behind tiny fibers. As with the heads, the key is to be gentle; some of these parts are easily bent.

FIG. 7–6 Clean everywhere the tape touches.

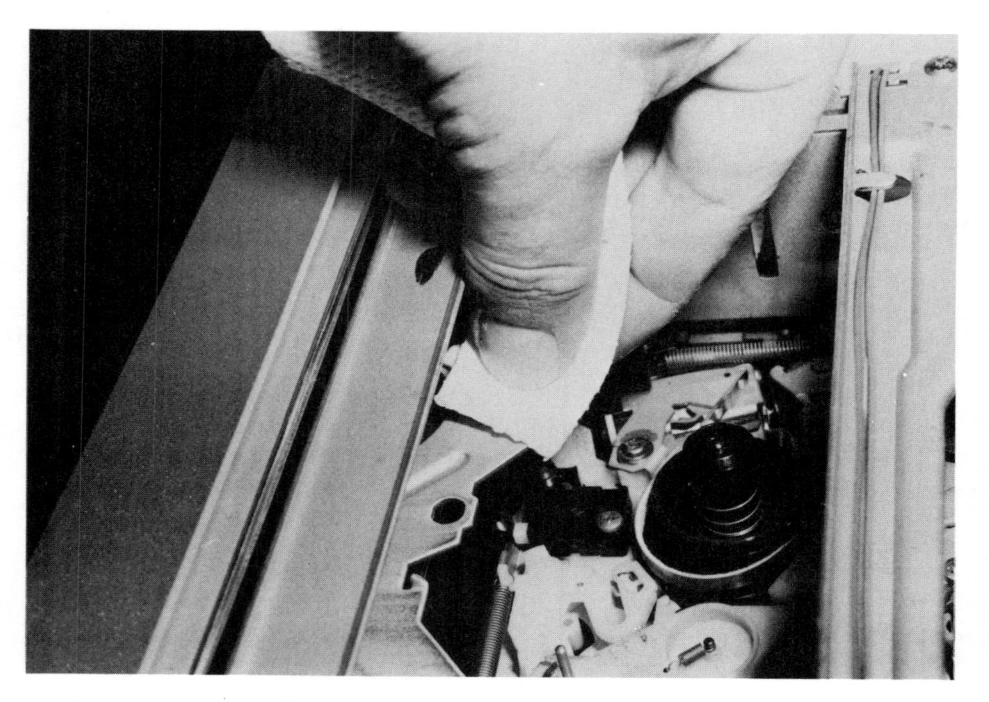

FIG. 7–7 Cleaning the cassette platform on a front-loading VCR.

One part of the VCR often ignored is the casssette platform. This is the flat plate or holding bracket(s) on which the cassette rests as it is loaded into the machine. Tiny pieces of plastic from the case can be scraped off; each time you load a cassette, these fragments get pushed deeper inside the unit—where they can do the most damage. Check and clean this platform regularly. If your VCR is one of the growing number of front-loading versions, you will have to open the cabinet to clean the platform.

DEMAGNETIZING THE HEADS

The basic principle behind recording and playback is magnetism. The tape itself is little more than a collection of billions of molecular-sized magnets glued onto a long strip of plastic. The heads react to the magnetic fields, or they create them. Along with these magnetic particles is the static electrical charge that is made whenever two substances are rubbed together—in this case, the tape and the various mechanisms of the VCR. After a while, the heads and other metal parts can take on a permanent magnetic charge.

Once the heads contain a magnetic charge, the tape can become damaged merely as it runs through the machine. The more you play the tape, the worse it gets. If the charge becomes severe, the image stored on the tape can be erased just as easily as if you ran a magnet along the tape.

The magnetic charge and static buildup will cause a gradual deterioration of the recording or playback. The image will be weak and erratic, and there may also be static on the audio section, poor color, or flagging and glitches. Allowed to continue, these flaws could become permanent parts of your tape.

Flagging is a breakup of the linearity of the reproduced image. This is particularly noticeable in the corners of the screen. If a telephone pole is a part of a scene and you notice that the top or bottom of the pole seems to be bending, the image is flagging. Glitches most often appear as lines or bars passing through a scene. Sometimes they may even be stationary, or nearly so.

The two most common causes of flagging and glitches are faults in the tape itself (low quality, overuse, flaking particles) or magnetized heads. You can prevent the first cause by buying only high-quality tapes and then taking good care of them. The second can be prevented by regularly demagnetizing the heads, either about once a month or whenever you open the cabinet.

An audio head demagnetizer should NEVER be used inside a VCR, even for the audio heads. An audio head demagnetizer has too strong a magnetic field for a VCR. If this powerful field gets near the delicate video heads (and you can't get near the audio heads without getting too close to the video heads), a physical vibration is set up. Almost without fail, this will cause the video heads to shatter.

Use *only* a head demagnetizer designed for the VCR (Figure 7–8). And use this for all heads and parts. No other demagnetizing tool should be brought anywhere near the VCR. (This also includes a degausing ring used for the television screen. If your television set needs to be degaused, be sure to move the VCR and all tapes several feet away.)

Using the head demagnetizer is easy. Shut off the power to the VCR and remove the cover. Hold the demagnetizer *at least* three feet away from the unit, then turn it on. Bring it gradually to the unit, but no closer than $\frac{1}{2}''$ to the head-drum assembly (Figure 7–9). (Don't allow the tool to touch anything.) With a lint-free glove on your hand, slowly rotate the head drum. Next, move the demagnetizer probe past the metal tape guides, the capstan(s) and any other metallic parts that come into contact with the tape (Figure 7–10). Finally, pull the demagnetizer away from the machine, just as slowly as you brought it near, until it is again at least three feet away, and turn it off. This simple

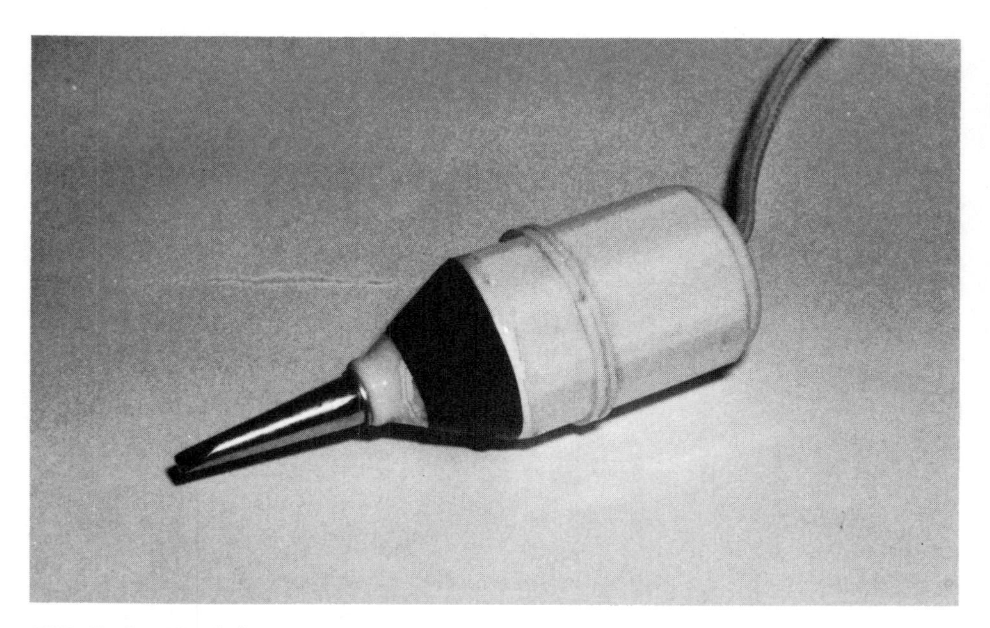

FIG. 7–8 Head demagnetizer.

procedure takes about two minutes, and it should be done each time you clean the heads. Your reward will be higher-quality recording and reproduction.

Be careful never to turn the demagnetizing tool on or off if the tool is closer than three feet from the machine. This could cause

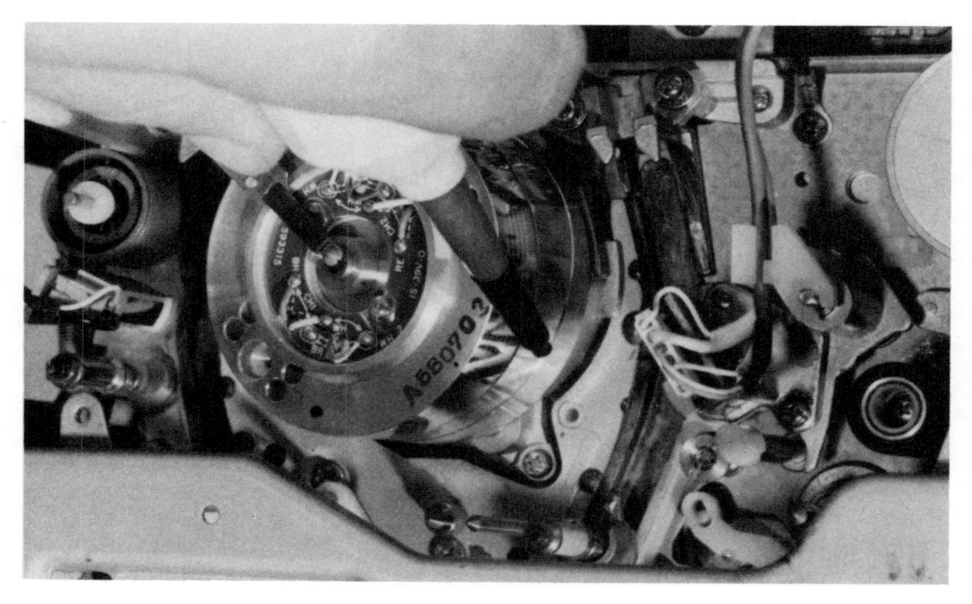

FIG. 7–9 Gradually bring in the demagnetizer and move it around the head drum.

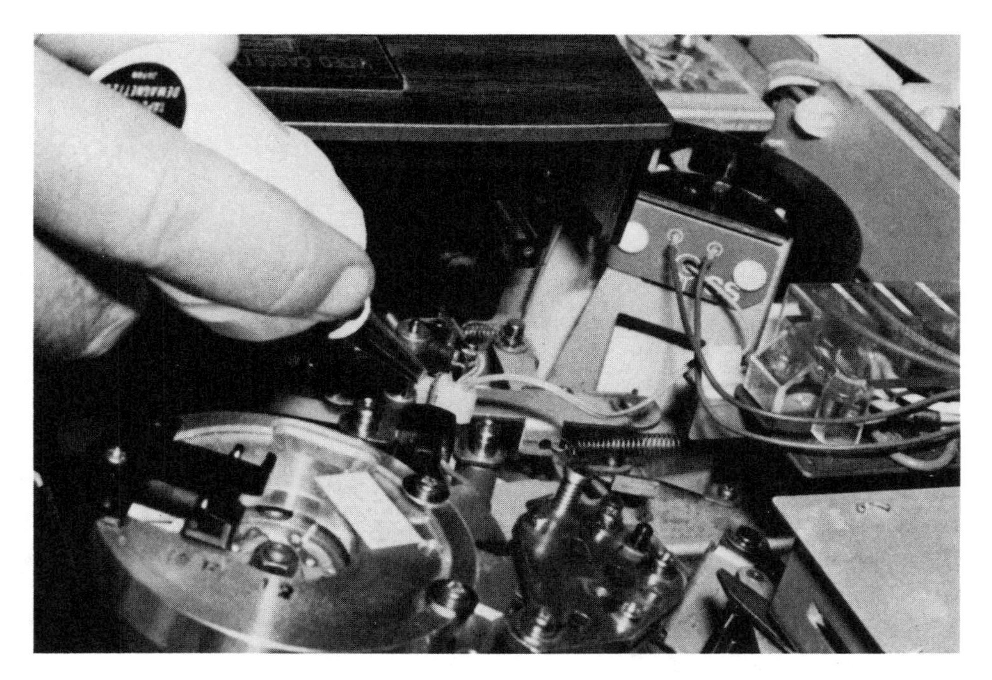

FIG. 7–10 Move the demagnetizer probe near other metallic parts of the VCR.

exactly the opposite of what you want to happen. Instead, the part will take on a permanent magnetic charge, and a stronger one than was present before you began.

One user followed the demagnetizing instructions to the letter. He carefully held the demagnetizer three feet away from the machine, with his arm fully extended behind him. He brought the demagnetizer into the machine, moved it around properly, and pulled away again. He did everything exactly right—except for one thing. Behind him, at just an arm's length, was his library of videotapes. The instant he flicked the switch, that magnetic field served as a sort of bulk eraser. His VCR was in great shape. He just didn't have anything left to play in it.

CHECKING THE BELTS

Each time you open the top of the machine for regular cleaning, take a quick look at the belts, pulleys, and other moving parts. Few machines allow access to the belts from the top (Figure 7-11), but this will at least make you aware that they exist in there somewhere.

After every 1,000 hours of operation, or every six months, remove the bottom cover for a more thorough examination. You're looking for

FIG. 7–11 Check the belts for wear.

worn spots, frays, or other obvious damage. A light touch with your finger will tell you if the tension and flexibility is still correct. Better yet, use a pencil so that you don't leave fingerprints.

If a belt shows signs of wear and tear, replace it *immediately*, before it has a chance to bind or break. If the belt is stretched and loose, you may be able to adjust it to keep it in service for a while longer. Since not all manufacturers allow for this kind of adjustment, check the service manual for your machine first.

Even if your VCR does allow the belt to be adjusted, it's still better to replace it. A stretched belt has a limited life span. Since you'll have to replace it sooner or later, you might as well do it while the machine is open. You'll save yourself time, money (the price of the replacement belt is *not* going to go down), and trouble, and your VCR will be in tiptop operating condition instead of just barely getting by.

In an emergency, such as on a weekend or holiday when you can't get a replacement, you can put a slipping belt back into temporary service by lightly applying some resin or beeswax to the inner surface (Figure 7–12). This will allow it to adhere better to the groove in the pulley.

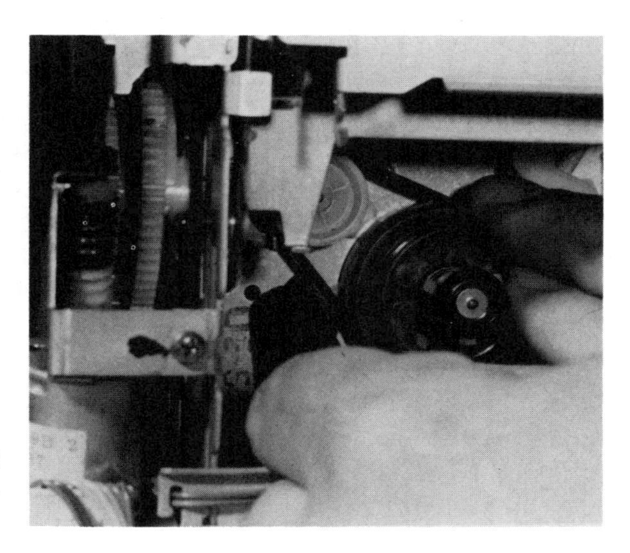

Fig. 7–12 Dressing a belt with resin is only a temporary solution.

Occasionally you might hear of someone who has successfully replaced a belt with a rubber band (we hope on a temporary basis). Although this *might* work, the hazards involved far outweigh the advantages. If the rubber band is too loose, it simply won't work. If it's too tight, it can yank and pull parts out of alignment. You might be able to finish watching that movie, but chances are good that the next day you'll be making an expensive visit to the shop for a complete overhaul.

Note: While you have the bottom cover off, carefully clean the underside of the machinery as well.

ADJUSTING TORQUE AND TENSION

Imagine a long silk thread being pulled from one person to another. If the first person releases the thread too quickly, there will be a tangled mess on the ground. If the second person pulls it in faster than the first can release it, the thread will snap. The tape inside a VCR is like this thread. The first person is the supply reel; the second person is the take-up reel. Both have to work in perfect unison. The tape isn't quite as delicate as a silk thread, but it comes close.

You will need two tools to measure and adjust this critical movement: a torque gauge and a tension spring. (Some tools measure both torque and tension.) A torque gauge tells you how much twist is applied by the motor, pulleys, belts, and related parts that transport the tape from one reel to the other. To take the measurement, gently press the tool onto the motor spindle.

The tension spring measures the tightness (or looseness) of the tape and the servomechanism springs. There are two basic types: fan and spring. The fan type contains a stick that moves across a scale as the stick is pulled. The spring type looks like one of the scales used by fishermen to measure a catch. With either tool, the tension is usually measured in grams.

Often the tools are available only through an authorized outlet of the VCR manufacturer. If there is no provision for adjustment, more extensive repair work may be required—namely, replacing the belts, pulleys, or drive gears.

The actual amount of torque and tension in a VCR depends on many variables. More often than not, each unit, even those made by the same company, requires a different amount. The only way you can find out what the numbers are for *your* machine is to look in the service manual for that particular make and model. Even the point, or points, where the measurements are taken differ from unit to unit. The service manual will also show you where (and how) to make the adjustments.

Don't try to adjust the torque or tension unless you have the exact numbers (such as 170 grams) and the proper measuring tools. Trying to make the adjustment by trial and error will invariably lead to a fatal error (in which case, turn to Chapter 6 to learn how to fix a broken tape or how to untangle one).

SUMMARY

The key to VCR maintenance is to prevent problems. The easiest and best way to do this is to keep the unit and all parts as clean as possible. You can't stop normal wear and tear, but you can reduce it.

Once every month or so (or after every ten movies), remove the top cover from the machine. This allows easy access to the most important parts of the unit—and the ones that get the most wear and abuse.

Clean and demagnetize the heads often. As long as you perform the tasks with care, you're unlikely to cause any damage. Professional recording studios often clean and demagnetize the heads on a daily basis. Letting a magnetic charge build up will harm the unit and shorten its life.

Clean all parts that the tape touches, and demagnetize all metal parts.

Clean all other surfaces as needed. The tape-loading platform

TABLE 7-1. Basic Maintenance Schedule

Daily (if VCR is used often)	300–500 hours (or every 6 months)
Clean outside of machine Remove any cassettes and check for rewind Organize and store cassettes properly Keep dust cover in place	Check belts Clean underside Check drive gears and pulleys Check springs, rollers, and capstans Adjust torque and tension Check adjustable voltages Change silica-gel packets Clean tuner (if mechanical) Adjust tape transport if possible (see service manual)
25–50 hours (or monthly)	***1,000 hours (or annually)***
Clean heads Clean tape transport mechanism Demagnetize heads Visually inspect for wear Check cables Clean contacts on plug-in boards Clean tape cassette cases and boxes	Professional checkup, including head alignment and other fine adjustments
	Occasionally
	Repack stored tapes before using Check power cord for signs of wear Reset timing devices

and brackets don't actually touch the tape, but they could pick up tiny particles from the tape case.

About twice a year, remove the bottom cover and check the belts, pulleys, and other parts that are hidden beneath the machine. Replace them as necessary. Clean the underside of the machine while the cover is off.

Torque- and tension-measuring tools can be used to check out the various motors and springs of your VCR. This can be done, however, only if you have a listing of the correct values. Since each machine is different, the only way to know for certain is to check the service manual for your make and model. NEVER try to make this adjustment with a "Well, that looks about right" attitude. This not only will not work, but it may seriously damage the VCR.

Table 7-1 lists basic maintenance procedures; how often you perform them will vary according to how often you use your VCR.

In Chapter 11, you'll find a handy log to use for your regular maintenance routine.

Chapter 8

The Electronics

Anyone capable of operating the controls of a VCR is also capable of taking basic measurements with a multimeter, or VOM. The VOM is handy for isolating the source of videotape difficulties, particularly in the power supply. The meter can also be used for other tests, on various circuit boards.

Often the test points and the values that should appear on the meter are printed or etched on the circuit boards. (Take a look the next time you remove the cover from your VCR.) Even if the measurements mean nothing to you, you can save time and money by presenting them to the technician. And by monitoring these various test points regularly (and recording the values read and corresponding locations at the back of this book), you can keep an eye on how the VCR is performing.

USING A VOM

Most of a VCR's video circuits, the critical record/playback head alignment, and the complex synchronization and color amplifiers/generators are too difficult for the amateur technician to handle. They also require expensive and complicated equipment to test or adjust. However, a VOM (Figure 8-1) can help you in many ways. For example, your entire set may be inoperative or a particular component may not work properly, with the difficulty being nothing more than a breakdown in the power supply or in some other part of the unit that can be easily found and fixed. One of the diodes that converts AC to DC may be damaged, a regulator or power transistor may have

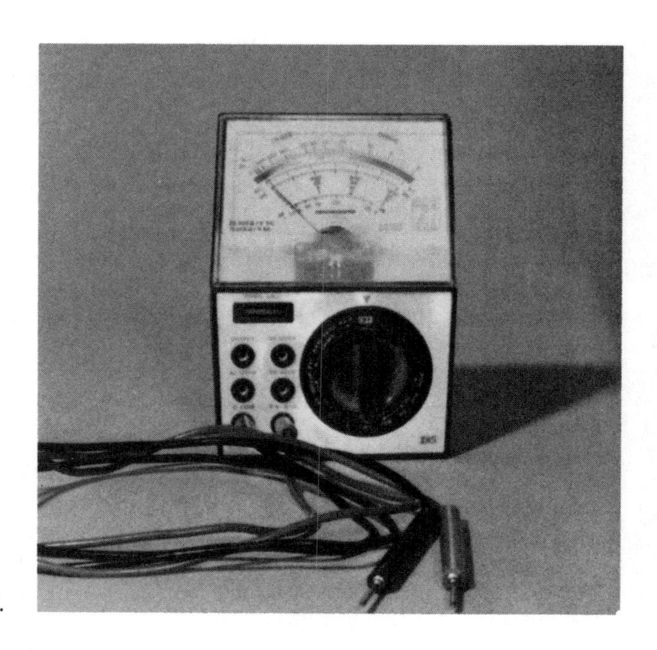

FIG. 8–1 A typical VOM.

opened or shorted, or something may be causing a short in the power supply, killing voltages throughout the set.

Using a VOM is simple. You can use it to check wall outlets, to check the wiring in your car, and of course to help you maintain and repair your VCR.

The two basic functions of the VOM are to read voltage and resistance. You can select the appropriate range for the voltage you'll be measuring.

Almost any meter will do for normal use. You don't have to spend hundreds of dollars on a fancy meter. The inexpensive $10 or $20 VOMs available will be accurate enough for your needs.

Voltage is measured as either AC (alternating current, such as what comes from the wall socket) or DC (direct current, such as the voltage from a battery or as used by most electronic circuits). Most meters have several ranges in each. Selecting the proper type and range is important. Try to read the 120 volts AC from the wall socket with the meter set to measure 3 volts DC and you're going to have some trouble. The same applies to setting the selector set at 3 volts DC when a 93-volt DC impulse is coming in.

A graphic example of what can happen concerned a new technician who was *sure* that he knew how to use the meter. In his haste, he set the meter to the $\frac{1}{2}$-volt range. The probes were touched to a 48-volt source. The needle instantly swung across the face of the meter and wrapped itself around the stop pin. His brand-new meter was ruined.

It is always best to start with a range setting higher than you think is necessary. For example, to probe a test point that you *think* should be 5 volts, don't just assume that you should place the VOM selector at 5 volts. You could be wrong.

The probes have metal tips, which means they can conduct electricity. Be extremely careful when you poke around live circuits (see Figure 8–2). It's easy to slip and create an accidental short. Even if the current doesn't harm you, that short circuit can destroy delicate components of the VCR. Always attach the probes with the power off. Double-check that they are where they should be and then energize the circuit. Turn off the power again before removing or moving the probes.

The other function of the VOM is to measure resistance, which is done in ohms. The ohmmeter side of the VOM can be used to test individual components (Figure 8–3). This chapter will tell you how to do this. The ohmmeter can also tell you if an inside wire or cable is broken. This second kind of check is called "testing for continuity." As an example, see the sections on how to make cables and connections in Chapter 5.

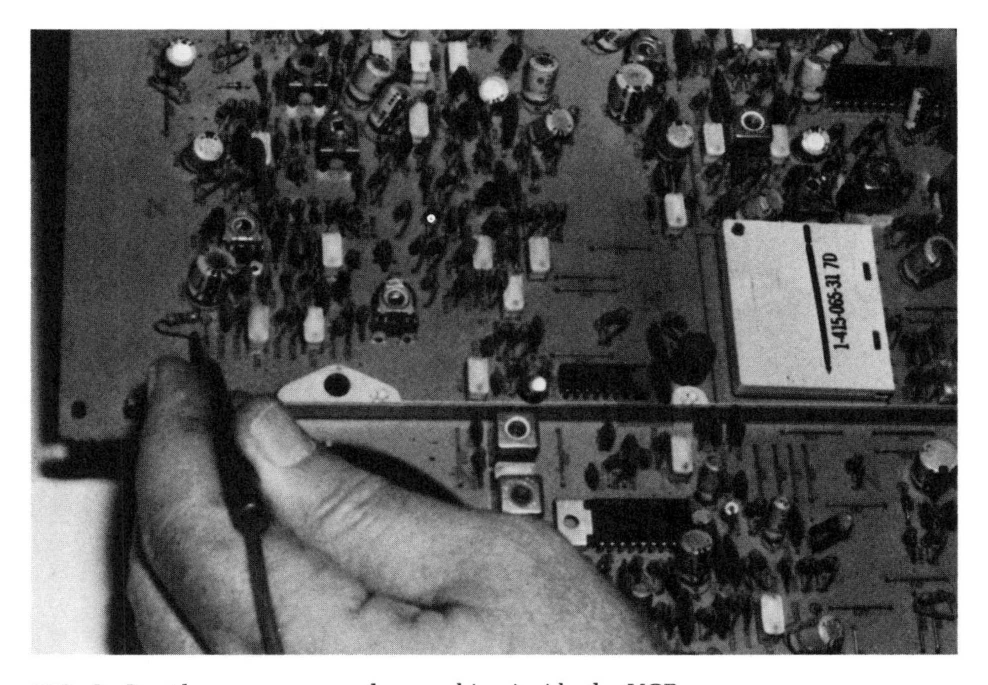

FIG. 8–2 Always use care when probing inside the VCR.

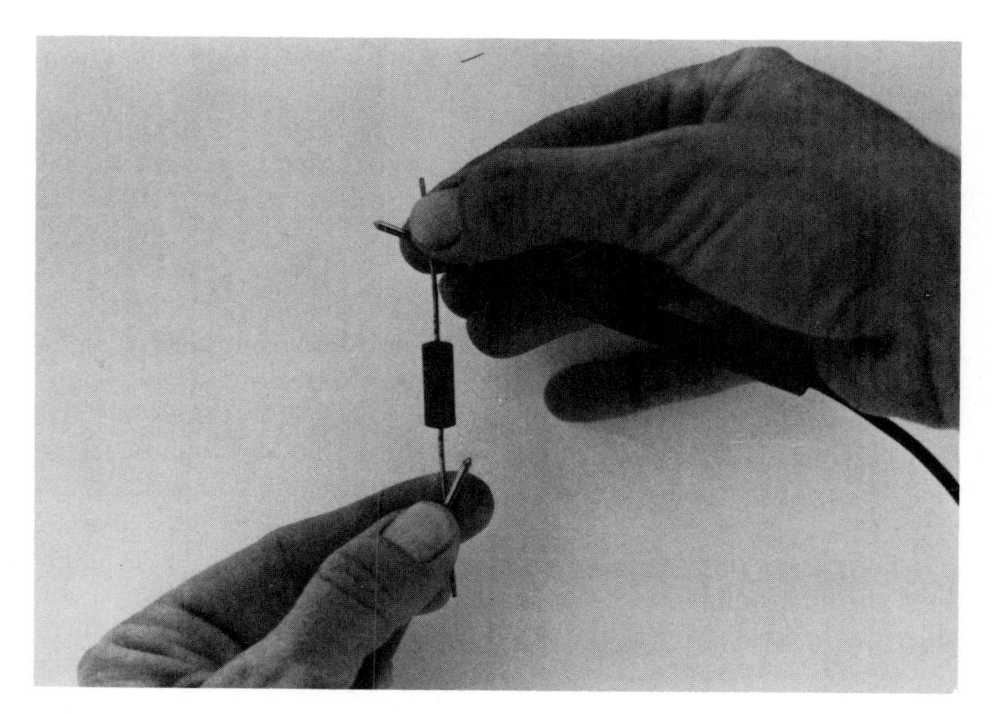

FIG. 8–3 Testing a component.

Some VOMs have a polarity switch (Figure 8-4), usually labeled "Forward" (or "Normal") and "Reverse." Since the DC voltages you'll be measuring can have either a positive or negative ground, a polarity switch comes in handy. If you see the needle move down, in the wrong direction, all you have to do is flip this switch.

FIG. 8–4 A polarity switch.

However, you can do the same thing merely by reversing the probes. The easiest way to do this is to unplug them from the meter and plug them back in again in the opposite holes. You can also reverse them on the spots being tested.

The two leads coming from the meter are usually black and red. Most of the time you will use the black probe as the common or ground (−) side, with the red (+) touching the spot being tested. Common spots may be marked (e.g., common, ground, G), and the chassis itself is common.

The probes are insulated for your protection. Even when you check a circuit with the power turned off, make it a habit to hold or touch the probes *only* by the insulated handles. This habit will serve you well: learn it early and you'll never have to learn the value of insulation the hard way.

THE POWER SUPPLY

The power supply of your machine (Figure 8–5) takes alternating current as supplied by a wall socket and converts it into the direct current required to operate the VCR. The 120 volts AC that comes into your home must be converted into appropriate voltages. This is done by a transformer, which takes the incoming 120 volts and steps it up or down to the values needed by the various electronic components of

FIG. 8–5 The power supply of the VCR.

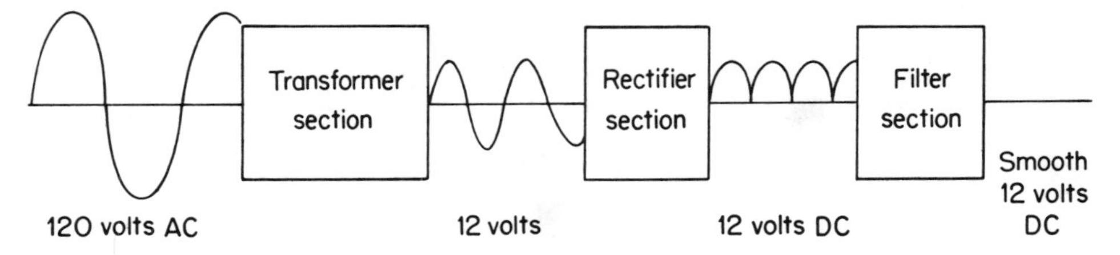

FIG. 8–6 The function of a power supply.

the set. Then rectifier diodes convert the current from AC to DC so that it passes in only one direction. A network of resistors and capacitors then filters the DC output by smoothing the AC sine wave to a steady DC flow. A regulator then takes over and maintains the output voltages at relatively constant values. Figure 8-6 shows a simplified version of how the power supply works.

The power supply does just what its name says: It supplies power to everything else in the machine. If the power supply goes down, so does the equipment. If part of the power supply fails, a circuit that takes power from that part will also stop functioning.

Checking the power supply is usually fairly simple, especially if your model's power supply uses plugs to apply power to the rest of the unit. Sometimes the values at these inputs are marked (Figure 8-7). Otherwise, check the service manual for your VCR. The section on the power supply will show you exactly what the output values are and where they are tested.

If neither is true of your machine, you can still test most of the electronic parts of the power supply with a VOM. Doing so will tell you whether the problem is in the power supply or in the circuits. If the fault is not with the power supply, the test will often tell you which

FIG. 8–7 The voltage values and test points may be marked on the circuit boards.

external stage to check next. Making the repair yourself can save you anywhere from a neighborhood shop's minimum service charge up to hundreds of dollars. Even if you can't locate the actual problem, at least you'll have eliminated some of the things that are *not* at fault. This in itself can save you time and money.

As you become familiar with your VCR and perform your first maintenance procedures, make it a point to search for these identified test points and values. Measure them, especially when the unit is operating properly, and note them in the space provided at the back of this book.

Even with a schematic or service manual, you may find that the actual voltages at the test points are slightly different from those stated in the instructions. Don't worry. If the unit is functioning properly, the values are obviously fine.

CHECKING THE POWER SUPPLY

If the VCR becomes inoperative, test the power supply with your VOM. If only some of the output voltages are available, the power supply may be faulty, or something inside the VCR may be causing one or more of the voltages to drop too low for you to be able to read accurately (e.g., a resistor shorts or breaks).

If the VCR is completely dead, the first and most obvious (but often overlooked) step is to check the wall outlet (Figure 8-8). Your VOM will tell you the exact voltage. If it's more than 15% out of range from 120 volts AC, the incoming power probably is insufficient.

Check the wall outlet to see if the VCR is plugged in. You wouldn't be the first person to start yelling about a broken machine, only to find that someone—perhaps even *you*—unplugged it.

If you know that power is getting to the VCR, you can assume that the problem is within the machine. Then you can begin to eliminate the possible causes of the failure.

The AC comes into the VCR and goes either to a switch and a fuse, or vice-versa. Check both to be sure that the problem isn't simply a bad switch or a blown fuse.

Note: Some VCRs use a special device called a *fusistor*. This is a two-prong component that is usually soldered into place in one side of the incoming power line. If you don't know what the fusistor looks like or where it is in your unit, refer to the service manual for your machine.

Whether testing a switch, a fuse, or a fusistor, the procedure is basically the same. At least one end must be disconnected from the

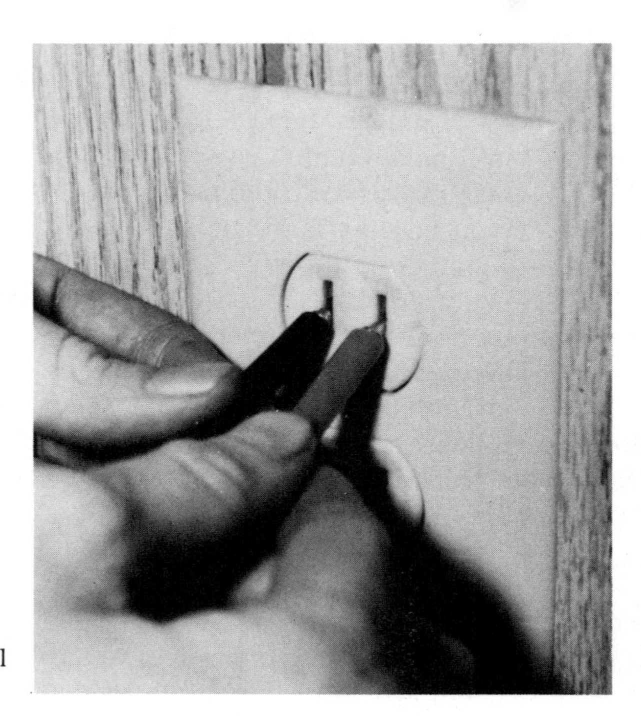

FIG. 8–8 Check the wall outlet for incoming power.

rest of the circuitry. (Do this with the power OFF and the unit UNPLUGGED.) Set the meter to read ohms (resistance) and touch the probes of the meter to the ends of the device. A fuse or fusistor will show a low resistance (50 ohms or less) if it is good. A switch should give a reading of almost no resistance with the switch in one position (ON) and infinite resistance in the other (OFF). Any other reading means that the device is bad and has to be replaced.

A basic step in determining if low voltages are the fault of the power supply or are a reflection of troubles elsewhere in the unit is to "unload" the power supply. If your unit's printed boards and servomechanisms use plugs and cables for the interconnection of the circuit boards, you can disconnect all outgoing branches in which the voltage seems to be incorrect.

Now take another reading of the output with the suspected circuit board disconnected from the power supply (Figure 8–9). If the trouble is outside the power supply, the voltage reading will return to normal or slightly higher. If that circuit board is okay and the trouble is in the power supply, the power supply will still give out the same wrong voltage level. Disconnecting the external feeds will then make no change in the measurements you get.

The same testing sequence can be done on each circuit. However, if too many of the outputs from the power supply show incorrect val-

FIG. 8–9 Testing the output of the power supply.

ues, the chances are good that your tests will simply show that you need to repair or replace the power supply.

TESTING COMPONENTS

Often a faulty component shows obvious signs of damage. A resistor might be cracked, or fluid might be leaking from a capacitor. Also look carefully for leads that have come loose and for broken solder joints.

We know of an owner who complained about a constant shift of color in her television/VCR system. Through a process of elimination, it was tracked to the color section of the television (which let the VCR off the hook). One technician she hired was ready to yank out the color module boards and replace them. At $85 each, these would have been a rather expensive way for the unwary owner to find out that she had chosen the wrong technician.

When we looked at the unit, we noticed that a transformer had become unsoldered and was just barely in its socket. Any small vibration caused the lead to break contact. A penny's worth of solder solved the problem.

Visual inspection is always the first step in diagnosing a malfunction. Not only can it reveal that the malfunction is actually something simple, but it can help you spot a damaged component.

Some of the components may be checked effectively with nothing more than a VOM. Diodes, capacitors, and many transistors, coils, and transformers can be checked with the resistance or continuity scale on the meter. Be aware, however, that most of the circuit boards in the VCR are delicate. You should limit most testing and all repairs on this level to the power supply. Leave the rest to the professional.

Most components can be checked with a VOM only when they are disconnected or removed from the circuitry. While they are still a part of the circuit, you will get readings of the overall circuit, not of the suspected component. In the case of diodes, capacitors, coils, and resistors, try the diagnostic tests described below with at least one end disconnected from the circuit.

A diode allows current to flow easily in one direction only, so it should have higher resistance in one direction than in the other. Put the range selector of the VOM on a high-resistance scale and attach a probe to each end of the diode. Some resistance value will be indicated. Now, reverse the probes, and the value should be considerably different, at least 2:1. If there is no difference in the two readings, or if the meter indicates no resistance (a dead short), the diode is bad.

Transistors are similar to two diodes connected back to back. You can check a transistor with a VOM for operating voltage while it remains in the circuit, but it must be removed from the circuit for a resistance check (Figure 8–10). You can also test a transistor with an in-circuit transistor-checking device (Figure 8–11).

There are three leads to a transistor: the emitter, the collector, and the base (see Figure 8–12.) Put the VOM probes on the leads between

FIG. 8–10 Testing a transistor with a VOM.

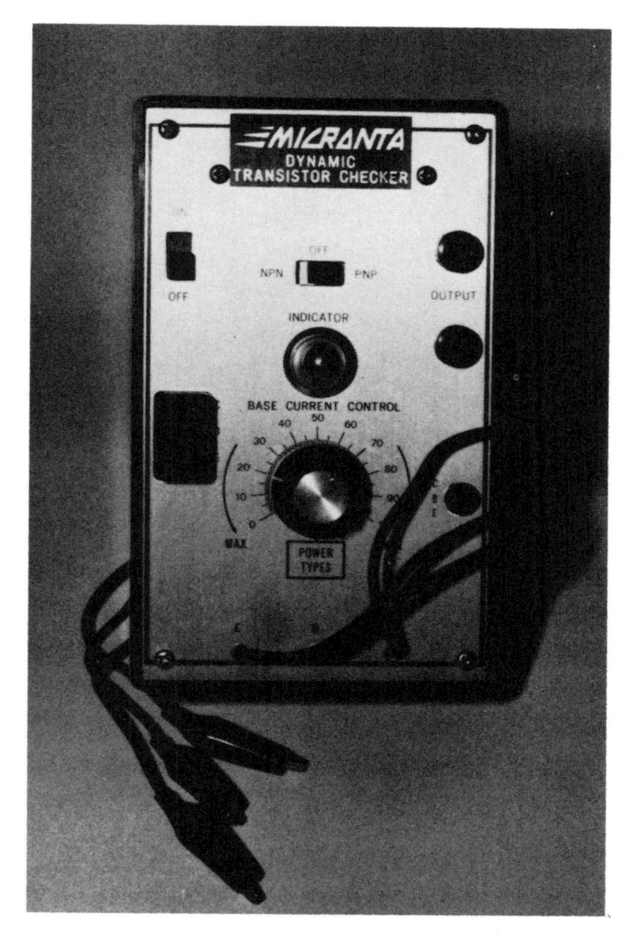

FIG. 8–11 A transistor tester.

the emitter and base and note the resistance reading on the meter. Reverse the leads (or throw the polarity switch). The indication should be the same as for a diode—high resistance in one direction and low in the other. If the two values are nearly the same or you have no resistance, the transistor is bad.

Similarly, use the leads to check the continuity between the collector and base of the transistor. Again, you should get high resistance in one direction and low resistance in the other. Normally, a high resistance value should register in both directions when the VOM leads are placed between the collector and the emitter.

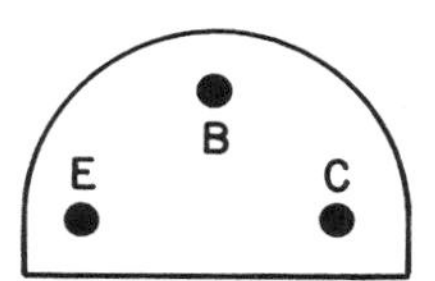

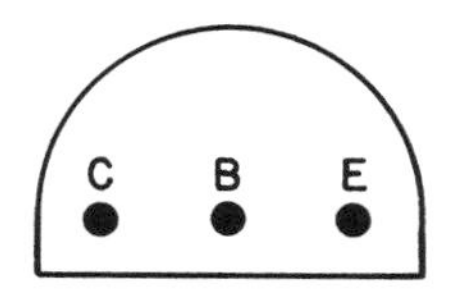

FIG. 8–12 The three leads of a typical transistor.

When it is inconvenient (or impossible) to remove a transistor from a circuit for testing, you can make some DC voltage measurements while the transistor is still in the circuit. The power must be turned on for this test, which means that it involves a degree of danger.

For most basic amplifier and voltage regulator circuits, the voltage should be between .15 and .7 volts directly between the emitter and the base. (You can also measure the voltage between emitter and ground, then between base and ground, and subtract the two, if you want to get complicated about it.)

If voltage is present, then test to see if you have collector current. To do this, take a voltage measurement from collector to ground. If there is no voltage, the transistor is probably bad. To be sure of the transistor's condition, use a transistor checker or remove the transistor from the circuit.

Whenever you use the probes of the meter, you must be particularly careful to avoid short-circuiting any points on the circuit boards. If this occurs, not only will your measurements be useless, but you can also severely damage the entire unit. Always shut off the power to the unit whenever you connect or disconnect the test leads, just in case you slip.

My coauthor and I worked on the preceding paragraph on a Saturday evening. On Monday morning, he was working on the power supply of a computer and apparently forgot what we'd just written. A wrench fell off the worktable, causing him to look away for a fraction of a second. A probe jiggled—and there went a $400 power supply. (Actually, we managed to fix the damage for much less, but the lesson should serve as a warning.) His only comment about the incident was, "Do as I say, not as I do—or did."

For checks involving test equipment more complex than a VOM, leave the testing and service to a professional.

REPLACING COMPONENTS

If you manage to locate a problem component, you can usually just replace it with a new part. Although you should be extremely wary of fiddling with the other circuit boards, replacing components on the less-sensitive boards and sections, such as the power supply, are fairly simple.

You don't have to be a trained electronics expert to replace a bad diode or transistor. Most components are plainly marked. Resistors, for example, have colored bands that tell you how many ohms that resistor provides, the percent of tolerance, and even how much power it will handle. Transistors and diodes are usually labeled with a code, such as "1N914." This will tell you exactly what you need for replacement.

Be warned that if you replace a bad component and the new one blows as well, the actual cause of the malfunction could be elsewhere in the circuit. You have several choices at this point. You can replace the *new* component, try to find what the *real* problem is, or take the unit to a professional and let him worry about it.

If you have trouble identifying the components or reading component values, take the bad part to an electronics supply store and ask for a replacement.

Any replacement has to be an *exact* match of the one removed. The wrong one simply won't work, or it may even cause damage to other circuits and parts.

As mentioned in Chapter 2, soldering is an art. Take the time to learn how to solder correctly *before* you replace any components.

MAKING ADJUSTMENTS

On some circuit boards you'll find tiny boxes with screws and a label along the side, such as "+5 Adjust." These adjustable variable resistors are called "pots." If the unit is working well, leave them alone. If your VOM indicates that the power coming into a board is incorrect, you can try to bring it to the correct value by carefully turning the screw.

Don't try to adjust any screw or device that is labeled "Bias" or *anything* other than "voltage." Adjusting these requires equipment much more sensitive and complex that a VOM,

SUMMARY

Testing and replacing the electronic components of your VCR are critical steps. Many of the circuits are extremely delicate. One mistake, and there goes a whole circuit board. Don't tackle the job unless you feel confident and know that you will exercise extreme caution. Testing and adjusting complex circuits requires complex equipment. These tasks are best left to the professional technician.

A VOM allows you to perform quite a few tests. The two most important tests are to find out if the power supply is providing the correct voltage to the boards, and, by testing for resistance, to find out if certain components are bad.

Diagnosis is a process of elimination. Using the VOM and some common sense, you can at least eliminate a few things that are *not* causing the problem.

Most of the time, you'll be dealing only with the power supply or with the rest of the VCR on a circuit-board level. Your tests should let you know quickly if the power supply is bad or which circuit board has gone down. You may have to replace the entire unit or board. (That's what they'd do in the shop for $35 to $60 an hour, plus parts.)

If you track the problem to a single component, you can replace the component. Never replace a part with one that isn't an exact match. And be very careful when soldering or desoldering. Don't do any soldering until you've thoroughly examined the circuits.

Sometimes a malfunction can be fixed by merely making an adjustment. Some adjustments can be made with nothing more than an insulated screwdriver and a VOM. If the correct value isn't printed on the circuit board next to the adjustment, refer to your service manual.

Chapter 9
Troubleshooting Guide

At the end of this chapter is a troubleshooting chart that lists some of the most common VCR malfunctions, their causes, and possible solutions. Your owner's manual and service manual may list additional problems and solutions for your make and model.

A good place to start learning how to troubleshoot is to pick up a book on basic electronics. You don't have to be an electronics engineer. An understanding of what a transistor does and how it does it, for example, will help you spot problems.

Some knowledge of physical mechanisms will help, also. For example, a large pulley drives a smaller pulley, and a belt connects the pulleys. The smaller pulley will spin faster than the larger one. (If the circumference of one is three times larger than that of the other, the smaller pulley will have to spin three times as quickly.) If the belt between the two breaks or begins to slip, the smaller pulley will not move properly, if at all. It's as simple as that.

Each VCR design has certain peculiarities and certain manufacturing weaknesses. For example, a guide pin in your particular machine might be prone to bending. Or a capacitor in the power supply might be known to give out at six-month intervals. Often, something that breaks down once will do so again. It might take you years to figure out all the little quirks of your machine and to learn what sorts of things to look for first inside.

TROUBLESHOOTING STEPS

One troubleshooting step is to talk with other owners. Another is to keep close track of what goes wrong. You can do this by keeping an accurate record in the Owner's Log and Owner's Notes in Chapter 11. Eventually the log may reveal a pattern. (If you sell your machine, give the new owner your maintenance log and explain its value.)

The easiest way to go about troubleshooting is to keep a reference list of possible problems, probable causes, and the steps for correction. The *best* way is to resist looking at the chart until you have examined the machine. First, try to find the cause of the malfunction. It might take time in the beginning, but you'll benefit in the long run by having a better understanding of how things work inside the VCR.

Troubleshooting Steps

1. ***Look for the obvious.***
2. ***Check for operator error (i.e., you).***
3. ***Check for cassette problems.***
4. ***Check for mechanical problems.***
5. ***Test the electronics.***
6. ***Use process of elimination.***
7. ***Get professional help (if the job is beyond you).***

CHECK THE OBVIOUS

Some years ago a major freeway was blocked by a large truck. The truck came to an overpass, but the bridge was about four inches too low for the truck to clear. The driver studied the problem, the sheriffs studied the problem, transportation experts studied the problem. Calculations were made on how long it would take either to remove a section of the bridge or to cut away the top of the truck. It seemed that the only sensible solution was to get all the traffic behind the truck off the freeway (several miles of vehicles, driving in reverse up the on-ramp and trying to disperse).

A young boy who was stuck in the traffic jam with his father was enjoying the whole scene. As eager as he was to watch a chunk of the bridge removed, he couldn't help making a suggestion. "Why not let some air out of the tires, drive under, and put the air back in?"

The moral of the story is that often a problem isn't a problem at all. Something simple has gone wrong—perhaps an insect has become caught in the mechanism or a belt is broken or slipping. Even more serious malfunctions often can be spotted just by looking.

The first step in proper diagnosis of an unknown problem is to make a visual check. This begins outside the cabinet. Is the unit plugged in? Is the wall outlet good? Are all the cables and external connections in the proper places, and are all the wires good? Is the problem ouside the VCR, perhaps with the television set?

If you can't find the problem and decide to go inside the cabinet, look for the obvious. What appears to be a serious malfunction could be nothing more than a broken or loose wire. A fuse may have blown. Bad components can often be spotted quickly. (If you find a burned component, don't forget to take the obvious step of wondering *why* that happened.)

By looking for the obvious, you could find that the actual cause is quite different from what *seems* to be wrong. This in itself can save you time and frustration.

CHECK FOR OPERATOR ERROR

Virtually all problems occur as a result of something the operator does (or doesn't do) somewhere along the line. This doesn't apply only to those operators who are new to video. Those who have been around for years make mistakes or overlook certain things.

"My tape won't load," could be something as simple as an upside-down cassette. (Sounds silly, but it happens.) A dubbing problem could be nothing more than incorrectly connected cables.

Read through the owner's manual that came with your VCR and learn how the various parts function. Before you blame the machine, be sure that you're not trying to get it to do something that it can't do.

CHECK THE CASSETTE

One of the weakest points of the VCR is the cassette. The tape can become tangled inside the machine. It can also break, twist, stretch, bind, or crinkle. The case mechanisms can cause so many problems that you'll swear that they were specifically designed to fall apart or otherwise ruin your viewing, if not your VCR. Beyond all this, the coating on the tape will eventually disintegrate.

Many malfunctions that appear to be a problem with the machine are actually nothing more than a bad tape, a bad cassette case, or both. Consequently, if something goes wrong, check out the tape before you suspect the VCR. There are obvious exceptions to this, of course. A flickering picture, for example, might lead you to believe that one or more of the circuits is giving out, In fact, the cause is more likely to be a flaw in the tape, from either manufacture or heavy use.

It's easy to check this. Simply try another tape. The two tapes may show the same symptoms, especially if they've been stored under the same conditions. If you suspect that the two tapes might be flawed, try a tape that's been stored under different conditions.

There is an exception to this. If the tape has been damaged while in use, the problem is more likely to be in the VCR. In this case, don't try another tape. You'll only ruin more cassettes and take the chance of further damaging the machine.

Before you give up, turn to Chapter 6 for more information on cassettes.

CHECK FOR MECHANICAL PROBLEMS

A close second to cassettes as a cause of trouble are the mechanical parts of the VCR. Anything mechanical is subject to wear. And the more complex the device, the more often it will present problems.

This brings us back to cleanliness. The more dust and dirt that accumulate on the moving parts, the more quickly the parts will fail. By keeping everything as clean as possible, you'll reduce the chance of malfunctions.

Most modern VCRs are manufactured with "sealed" motors and solenoids designed to work for many years without lubrication. However, after a year or two of operation, accumulations of dust or moisture can cause these devices to freeze—and the VCR to malfunction or quit. Professional technicians often experimentally lubricate and move a malfunctioning device.

A drop of oil placed on the motor shaft *(with the power off)* might restore a frozen motor to operating condition, even if that motor is not meant to be lubricated (Figure 9–1).

Use extreme care whenever you try to lubricate the VCR unit. Certain mechanisms—servomechanisms and solenoids (spots where metal contacts metal, or metal contacts plastic)—may function better with a very light application of machine oil. This oil, however, is extremely damaging to videotapes and heads. Oil should not come in contact with videotapes, tape guides, or the rotating video heads. If your hand is unsteady or the nozzle of the oil container allows any more than the tiniest drop to come out, don't attempt any lubrication.

If oil spills on a critical area of the machine, clean it up with a cotton swab or one of the specially manufactured head-cleaning pads,

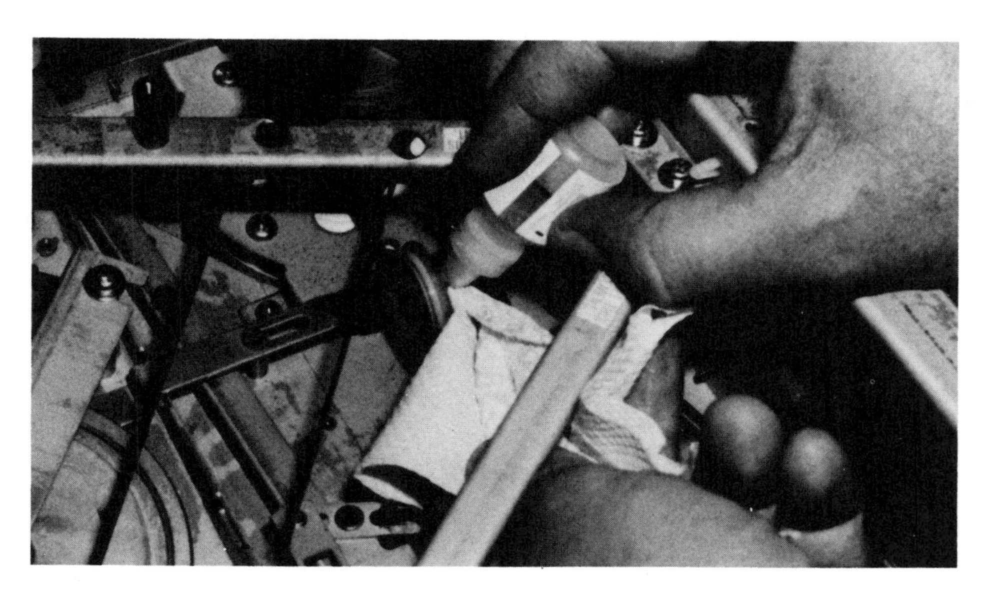

FIG. 9–1 Lubricating a moving part.

along with either head-cleaning fluid or pure isopropyl alcohol (see Chapter 7 for more details on cleaning the heads).

During cleaning and whenever you have the chance, inspect all moving parts for wear. The rollers may have developed flat spots. Often this means that the rollers are not rotating correctly on the spindles. It's rarely a good idea to attempt to lubricate the spindles, since even the tiniest drop of oil can permanently contaminate that roller (which will in turn contaminate any tape you use) and the machine itself.

A front-loading VCR has an additional mechanism, the set of parts that pulls the cassette into the machine. This assembly consists of a motor, belts, gears, levers, and other parts that are all finely tuned and balanced to operate in conjunction with the other parts. Disassembly and repair is complicated—virtually impossible without a service manual for your specific make and model—and brings with it the distinct danger that you will never get everything back together properly. Normally, these things should not be touched by the owner. If a visual examination and a gentle cleaning don't take care of the problem of loading or ejection, it's best to leave the task to a professional, one willing to accept responsibility for any damage done.

Have the service manual close at hand as you attempt to repair or replace any parts. Proceed slowly and carefully. You can cause a lot of damage by being in a hurry.

TEST THE ELECTRONICS

Chapter 8 tells you how to test the power supply (which in turn tests the other circuits) and the electronic components. This isn't difficult to do. The results of the tests will usually point to the cause of the problem if it is electronic in nature. The testing detailed in this book is necessarily basic. The specifics you need are found on the circuit boards and in the service manual for your unit.

Keep in mind that even though the tests are simple, there are dangers involved. The 120-volt AC is the most dangerous thing as far as you are concerned. Fortunately, there aren't many "hot spots" inside the VCR. However, there are a number of places where a slight error can cause instant and expensive damage. Before you do anything, read Chapter 2 again on safety precautions. And always remember the single, best rule: You can never be *too* safe.

USE THE PROCESS OF ELIMINATION

As you perform the troubleshooting steps, you'll get closer and closer to the cause of the trouble by the process of elimination. For example, nothing happens when you flip on the power. The fault is either inside or outside. Checking the wall outlet and the power cord tells you that the problem is in the unit. Once inside, check the fuse and the switch. If both are good, then the problem is either in the power supply, in one of the circuit boards, or in the wiring.

Faulty wiring often can be determined through a visual check. This now leaves only the power supply or the circuit boards.

So, out comes the VOM. By following the steps given in Chapter 8, you'll find out whether or not the power supply is functioning properly. If it isn't producing the needed voltages, you've isolated the malfunction to the power supply. If it *is* supplying all the proper voltages, you'll still know where the problem is, since the tests will show you which circuit board is causing the power supply to diminish.

Once you've eliminated all the sections that are *not* at fault, you can concentrate on the bad section. A visual check may reveal that a particular component is damaged. A schematic of the section will help you to track down a faulty part that doesn't show. Again, you're eliminating the good things, one by one, until only the bad is left.

If you don't feel confident that you can find a malfunction on the component level, you'll still have saved a great deal of time and money. Instead of bringing in a malfunctioning unit with an "It's broken—fix it," you'll be able to say, "The power-supply regulator board is acting up."

Many of the electronic repairs, especially those outside the power supply, can be handled by simply swapping a board. Even if you have a technician do the actual swapping, being able to tell him exactly which board has failed will turn the job from an expensive risk to a quick, inexpensive (relatively) swap.

GET PROFESSIONAL HELP

When all of your best efforts fail, turn the job over to a professional. Attempting to do a job when you lack the skill, knowledge, and equipment will only make things worse.

There are certain things that you should never attempt. If your diagnosis shows that the heads need to be replaced, it's time to bring the machine into the shop. It's usually wise to leave all repairs of circuit boards to a professional.

TROUBLESHOOTING CHART

Symptoms	*Cause*	*What to Do*
Nothing happens	No power	Check outlet and cord Check fuse Check power supply
No video, or no audio	Blank tape	Try another tape
	Bad connections	Check cables and connectors
	Dirty or bad heads	Clean or replace
	Bad speaker (audio)	Replace
	Bad circuit board	Replace
	TV adjustments off	Adjust or fix
No color	Tracking is off	Adjust
	Television isn't adjusted	Adjust
	Dirty heads	Clean
	Bad color board	Adjust or replace
Wiggly picture	Tracking is off	Adjust
	Dirty heads	Clean
	Dirty transport	Clean
	Tension off	Adjust
	Bad tape	Try another
	Bad horizontal control	Adjust or replace
	Servomechanism problems	Adjust or replace

Continued on next page

TROUBLESHOOTING CHART—*continued*

Symptoms	*Cause*	*What to Do*
Excessive dropouts	Bad tape	Try another
	Dirty head	Clean
	Tracking is off	Adjust
	Dirty tuner	Clean or replace
No recording	No input	Check cables and connectors
		Check instructions
	Stuck relay	Try "Record" several times
Poor recording	Poor input	Check and correct
	Dirty heads	Clean
	Bad erase head	Check and/or replace
	Bad tape	Replace
Playback but no recording	No signal	Check antenna or cable or related parts
	Dirty head	Clean
	Misaligned head	Align
Faulty playback, no playback, tape won't load	Belts	Check and replace
	Bad tape or case	Try another
No capstan, roller, head, or motor motion	Power	Check wall outlet, cord, fuse, and power supply
Switch malfunction	Bad switch	Replace
	Bad linkage	Replace
Tape binding, or sticking	Bad tape	Try another
	Transport trouble	Clean or replace worn parts
	Excessive moisture	

Chapter 10

When to Get Professional help

No matter how well you maintain your VCR or how much you learn about repairing it, sometimes you will have no choice but to call in a professional (and pay those professional fees!). Certain repairs require special, expensive equipment. Others require special knowledge that is far beyond the scope of any single book.

This book is meant to reduce to a minimum those times when you have to hire a professional and to cut the amount you'll have to spend when those times occur. When you consult a professional, you will already have taken care of many of the steps of diagnosis and will be able to supply a great deal of information to the technician. Since you've spent the time, he or she doesn't have to, and you won't have to pay for the time.

This book can go a long way toward reducing costs, but it can't eliminate them altogether.

FINDING A TECHNICIAN

It is often difficult and frustrating to find a reliable technician in *any* field. This isn't to say that there aren't plenty of competent and honest repairpeople or that you should automatically suspect every one of them. Just be cautious, and don't accept everything you're told.

Few shops can handle every kind of repair on all VCRs. For complex jobs, such as aligning heads, the technician should have experience specific to your make and model.

There are no federal regulations governing electronics repair. A license is needed only when the technician is working on transmitters. However, most states have some form of industry self-regulation. Certification is given to those who pass a test, such as one of the Certified Electronics Technician (CET) exams. This is a written exam, not a hands-on, "Can you really fix that machine?" test. All the certificate says is that the person had enough technical information tucked inside the brain to pass the exam. It says nothing about manual ability or about honesty and integrity.

Exercise the same commonsense precautions in seeking a good electronics technician as you would in getting a qualified auto mechanic or doctor. The best way is to check with friends, or with other owners of VCRs like yours. (And if you find a good technician, share that information with others.) The best recommendations will come from consumers, not from the sales staff of a store that has a service bay in the back. Also try to talk with the serviceperson who will be working on your machine. This will provide you with a fair indication of his or her competence and attitude. Don't be afraid to ask if the technician has had experience on your particular model, especially if the job is touchy.

A technician who is rude or refuses to talk about the required repairs should serve as a large, red danger signal. If he or she has this attitude before you pay, it will probably not improve after your check has been cashed and the problem still exists.

Price alone is not a good criterion for picking a serviceperson or shop. The apparent costs can be misleading in both directions. High charges don't necessarily mean high quality. At the same time, economy rates could, in the end, be extremely expensive. Contrast a low-priced but inexperienced technician who charges $10 an hour and spends four hours on a job and a technician who charges $35 an hour and gets the job done in less than an hour. The one who charges more may charge less in the end.

You may be lucky enough to find the best person for the job in the first shop you see in the Yellow Pages or on the street. You may also find someone who will charge you for repairs that were not necessary or will claim to have replaced parts that were never touched.

TERMS OF REPAIR

To help keep things honest, get a warranty that the work will be done to your satisfaction. A $40 bargain repair is useless if the serviceperson won't stand behind the work.

Get everything in writing, including the cost estimate for the repair. This estimate might be slightly different from the actual cost, but if the difference is major, you most likely have legal recourse. (To further protect yourself, have it written on the estimate that the cost is not to exceed a particular amount without your permission.) Along with the quoted price, the estimate sheet should note exactly what is to be done and when, and it should state the warranty terms. The warranty on both parts and labor should be at least 30 days.

When the work is complete, ask for an itemized list of what was done and the cost of each item. (Jot down this information in the space provided at the end of this book.) This is your protection for any warranty service and is also good for future reference. To avoid complications, request the itemized list *before* you leave the VCR. Some shops automatically keep itemized lists; others do not.

Before any work is done, let the serviceperson know that you want all the old parts that he or she replaces (unless they have trade-in value). If you give the serviceperson the idea that you know what is inside the machine, he or she will be much less likely to try to tell you that a burned out RCA video head came from your Hitachi. Keep the old part and check it against the replacement the next time you open the machine. If the shop refuses to give you the parts, go elsewhere.

With the information in this book, you should be able to provide a considerable amount of information to the technician. Your goals are to reduce the cost of repair and the amount of time that repair will take. Be sure that the technician knows what is wrong and what the symptoms are. The more information you can provide, the better.

DEALER RESPONSIBILITIES

When selling you equipment, the dealer assumes a certain amount of responsibility. (If the dealer doesn't, you should probably find another.) The dealer should make sure that the VCR is functioning when you buy it. If you buy an entire system, the dealer should see to it that everything functions as a unit before turning it over to you. If the dealer is content just to hand you a pile of boxes, you might as well buy through a mail-order company and save some money. Don't be afraid to ask the dealer to put promises in writing. Know where you stand and what your options are.

My coauthor bought a new VCR not long ago. It was on sale and was just the addition he needed for his system. Although he didn't need technical help to make the connections, curiosity got the better of him. He explained what he had in mind (a multiple television hookup) and asked, "How do I connect everything?" The salesman got out a piece of paper and drew sketches of two ways to cable that particular scheme. He then made sure that my coauthor had everything necessary and that he understood the instructions. Then he wrote his name and phone number on the slip of paper. Needless to say, he not only made the sale, but we now recommend him to other potential customers. He showed that important sense of responsibility to the customer.

After you make the purchase, the dealer continues to have the responsibility of customer care. If you have a problem a few weeks or months later, you should feel welcome to call in with questions. Service during the warranty period is obvious, but the service of answering questions should continue beyond this period.

Manufacturers are famous for refusing to talk to the customer, so the responsibility usually falls on the dealer to provide customer service.

The dealer should provide competent technical assistance after the sale on the basic operation of the VCR, as well as skillful repair when something goes wrong.

Small dealers and department stores do not have a technical staff. If this is the case, the dealer should be able to guide you to the appropriate servicepeople.

THE SERVICE MANUAL

The operator's manual that comes with the VCR does not usually list the routine maintenance schedule and precautions. Instead, many makers of electronic equipment try to encourage the buyer to purchase a service contract from that company's dealer at the time of purchase to make sure that maintenance is handled by the dealer.

The routine maintenance steps needed to keep your VCR in peak condition are described in more detail in the service manuals for each particular make and model of VCR.

There are many brands of VCRs, but only a few manufacturers. This makes things both easier and more difficult when trying to find the appropriate service manual for your make and model. The name on the front of your VCR doesn't necessarily mean that it's the source for the service manual. For example, if you own a Magnavox VCR, Magnavox itself may not have the service manual. This machine is actually distributed by North American Phillips (which also handles Philco and Sylvania).

Sometimes you can track down a service manual for your machine by going directly to the manufacturer. Other times you'll come across a manufacturer that either doesn't actually make the machine (and thus doesn't have the manual) or simply doesn't deal with the general public. The best and easiest way to obtain the service manual is when you buy the VCR. Be sure to specify the *service* manual. The user's manual is packed with the unit. The service manual is not.

Sometimes the service manuals are available only to manufacturer-authorized dealers. But usually the dealer will order a copy for you. The cost of the service manual is usually between $25 and $50.

Before plunking down your dollars, be absolutely sure that the salesperson writes on the receipt that the purchase price includes the service manual. That obligates the shop to find one for you or refund your money.

If you buy a used machine, you may have to dig a little deeper to find the service manual. The first step is to contact a local dealer or distributor of that particular make. Lacking that, look in the user's manual for the address or phone number of the manufacturer.

The names and addresses of the major VCR manufacturers will help you track down the service manual for your make and model. The following listing is far from complete, since there are hundreds of lesser known brand names. It is a general guide only—to give you a starting place.

MAJOR MANUFACTURERS

Aiwa America, Inc.
35 Oxford Drive., Moonachie, NJ 07074

Akai America, Ltd
800 West Artesia Boulevard, Compton, CA 90224

Curtis Mathes
1 Curtis Mathes Parkway, Athens, TX 75751

Fischer
21314 Lassen Street, Chatsworth, CA 91311

Hitachi Sales Corporation of America
1200 Wall Street West, Lyndhurst, NJ 07071

JVC
41 Slater Drive, Elmwood Park, NJ 07407

Magnavox
NAP Consumer Electronics, I-40 & Straw Plains Pike, Knoxville, TN 37914

MGA/Mitsubishi
3030 East Victoria Street, Compton, CA 90221

NEC
1401 Estes Avenue, Elk Grove Village, IL 60007
Panasonic
50 Meadowland Parkway, Secaucus, NJ 07094
Philco
NAP Consumer Electronics, I-40 & Straw Plains Pike, Knoxville, TN 37914
Quasar Electronics
9401 West Grand, Franklin Park, IL 60131
RCA Consumer Electronics
Building 1-400, 600 North Sherman Drive. Indianapolis, IN 46201
Sampo Corporation of America
1050 Arthur Avenue, Elk Grove, IL 60007
Sansui Electronics
1250 Valley Brook Avenue, Lyndhurst, NJ 07071
Sanyo Electric
1200 West Artesia Boulevard, Compton, CA 90220
Sharp Electronics Corporation
10 Keystone Place, Paramus, NJ 07652
Sony Corporation
700 West Artesia Boulevard, Compton, CA 90220
Sylvania
NAP Consumer Electronics, I-40 & Straw Plains Pike, Knoxville, TN 37914
Toshiba America, Inc.
82 Totowa Road, Wayne, NJ 07470
Zenith
1000 Milwaukee Avenue, Glenview, IL 60025

SUMMARY

You have to decide whether or not you can handle a repair job that comes up. If you have doubts about doing the job yourself, go to your local dealer or repair shop.

Both the dealer and you have responsibilities. The dealer owes the customer all the necessary support and should be willing to stand behind its products. The staff should be competent enough to give sensible advice as to which products will best suit your needs.

Communication is one of the things to look for when finding a serviceperson. If the technician isn't willing to talk to you before doing the job, he or she will be even less likely to talk to a dissatisfied customer.

Before you leave your VCR for repair, get a written agreement stating what is wrong, how much the repair will cost, the maximum you are willing to pay if the estimate is incorrect, the repair time required, and the conditions of the warranty. Always ask for the old parts that have been replaced and that do not have trade-in value (And be sure that if there is trade-in value, it is written in the receipt.)

If you plan to carry out more than simple routine maintenance, invest in the service manual for your make and model.

Know your rights and options. Then use common sense. Usually it's as simple as that.

Chapter 11

Maintenance Log

Use this chapter to keep track of what you've done and when you've done it. Dating your repairs and parts replacement is important, especially if you intend to remove the cover only rarely and plan to perform the head-cleaning chores with a cleaning cassette.

At the end of the maintenance log is a section for recording miscellaneous things that you've had to do or have done. For example, if a head has to be replaced or a belt changed jot down the date and the cost. Also keep a record of voltage and wire color codes.

This is also a good place to jot down operational peculiarities that you noticed, either with the machine, peripheral equipment (cameras, editors, rewinders), or certain tapes that seem to be performing differently from others in your library.

BASIC MAINTENANCE SCHEDULE

25–50 hours (or monthly)	*Date*								
Clean heads									
Clean tape transport mechanism									
Demagnetize heads									
Visually inspect for wear									
Check cables									
Clean contacts on plug-in boards									
Clean tape cassette cases and boxes									

continued on next page

BASIC MAINTENANCE SCHEDULE—*continued*

300–500 hours (or every 6 months)

Check belts									
Clean underside									
Check drive gears and pulleys									
Check springs, rollers, and capstans									
Adjust torque and tension									
Check adjustable voltages									
Change silica-gel packets									
Clean tuner (if mechanical)									
Adjust tape transport (if possible—see service manual)									

1,000 hours (or annually)

Have heads aligned									
Professional checkup									

Occasionally

Repack stored tapes before playing									
Check power cord for fraying									
Reset timing devices									

***Note:* This maintenance log assumes that you are following a routine schedule of cleaning. Some of the steps will necessarily be skipped if you are using just a cleaning cassette. In this case, perform the steps whenever you have the cover off.**

Owner's Log (Miscellaneous Steps)

Action	*Date*	*Comments*

Owner's Log (Miscellaneous Steps)—*continued*

Action	*Date*	*Comments*

Owner's Log **(Miscellaneous Steps)—*continued***

Action	*Date*	*Comments*

Owner's Log (Miscellaneous Steps)—*continued*

Action	*Date*	*Comments*

Owner's Log (Miscellaneous Steps)—*continued*

Action	Date	Comments

Owner's Notes

Subject ______________________ *Date* ______________

Subject ______________________ *Date* ______________

Owner's Notes—*continued*

Subject ______________________ *Date* ______________

Subject ________________ *Date* ________

Owner's Notes—*continued*

Subject ______ *Date* ______

Appendix
Cameras and Other Equipment

If you don't already own a camera or some of the the available video recording accessories, it probably won't be long before you'll want some such equipment. If you've been making home movies with a standard camera, you're in for a pleasant surprise the first time you use a video camera.

A video camera can easily cost twice as much as the VCR. If you dig out the camera once a year to make a two- or three-minute taping of opening Christmas presents, then owning a video camera may not be for you. However, if you enjoy making home movies but make very few because of the high cost of movie film, then the purchase of a camera may be one of the wisest investments you'll ever make.

Even if you use your VCR only for taping programs off the air or playing rented movies, you may want to buy various pieces of peripheral equipment: a rewinder to save wear and tear on your VCR's rewind motors, a video enhancer to improve the image of an aging tape and help in dubbing, or special locks and alarms to protect your investment.

Unfortunately, most cameras and peripherals are built so that the owner is likely to be able to perform only a very few minor repairs. For the most part, you'll be confined to tracking the problem to a particular piece of equipment or an individual component. Unless you have the necessary tools and experience, attempting to carry out major repairs is more likely to create new (and expensive) problems. In addition, it is difficult to find a source for parts.

CAMERAS

Until you've worked with a video camera, it's difficult to understand just how nice they are. Once you've used one, you'll wonder why you settled for that film camera all these years.

FIG. A–1 RCA color video camera.

FIG. A–2 Cannon color video camera.

FIG. A–3 Omnipro color video camera with built-in character generator.

The major drawback of the video camera is its cost, which can be several times the cost of the VCR. The better the instrument, the more expensive it will be. This means that care in choosing and maintaining a camera is important.

Before you purchase a camera, test it for possible malfunctions and make sure that it will do what you need it to do. If the camera is in good shape when you buy it, it will probably stay that way for many years unless you drop it or have some other kind of accident. Actual malfunctions are rare.

Once you have the camera, take care of it. Keep it clean and store it in a good-quality bag or other container. NEVER aim the camera at a bright light, and avoid aiming it at still objects for extended periods of time.

BUYING A CAMERA

Determining which camera to buy is similar to deciding which VCR to buy. Often it begins by coming up with what you want to pay. Cam-

FIG. A–4 Close-up of Omnipro camera with built-in character generator.

eras are generally priced in the $500-to-$1,000 range for a basic home unit. You can find cameras for less, and you can easily find ones for more.

The largest factor in the price of a camera is the features it offers. Although virtually all cameras at any price have a "white balance" control and a focus, the more expensive cameras also offer hue and color controls, character generators, faders, and other features.

Even if you know exactly which brand and model you want, insist on testing the camera before you plunk down your money. Basically this is little more than seeing the camera in operation and perhaps making a quick test tape. You can find out just about everything you need to know with a few simple tests.

With the camera connected to a monitor, look at the image on the screen. The colors should be clean and the image sharp. Don't expect the quality you normally see on television, though. The cameras used in broadcasting and commercial taping cost thousands of dollars—and the programs you see have often used complicated lighting to provide high-quality reproduction.

If the colors aren't sharp, the problem may simply be with the white balance. You can find out by focusing the camera on a white surface—preferably one that covers 100% of the view—and pushing

FIG. A–5 Close-up of Cannon camera control.

the "white balance" control. If the camera has hue and color controls, try adjusting these as well. If the reproduction is still bad, you probably don't want that particular camera.

Another simple test is to move the camera while it's in operation. What you see on the screen should be very close to what your eye would see moving across the same scene. Look for smearing and traces in the image. For example, once the camera moves away from a table or chair, an image of it should not lag.

This lag will be more obvious under low-light conditions. Only a very expensive camera will have no smearing under low light; the cheaper the camera, the more smearing it will have. Smearing under normal light (e.g., full sunlight) indicates a tube that is either aged or bad—and a camera you don't want.

Even more important than lagging is the problem of colored shadowing. Video cameras use the primary color green as the base color. Red and blue mix in over the green. The blending creates all the other colors you see. The balance of red and blue to green is critical. If the alignment is off, an image may have a red or blue shadow—a small blur to the side(s). Realignment of a camera is an expensive process—something very few outside the factory can do successfully. Therefore, don't even consider buying a camera with an alignment problem.

Also be sure to test the features offered on the camera. The salesperson should be able to explain these to you, which makes the test doubly valuable in that you'll get a bit of free training. You can test most features with nothing more than a camera and a monitor. (You can see the camera's quality simply by connecting it to a monitor.) Testing on an actual tape is better yet.

A final caution involves compatibility. Although new technology makes incompatibility rare, it *can* happen. Even if the salesperson is certain that the camera will work with your machine, get that promise *in writing*. You should have the option to bring the camera back for a full refund, or at least for an exchange.

SPECIAL CAMERAS

The trend in video equipment is toward single units containing both camera and recorder. Called "camcorders," "Betacams," and a number of other names, these units offer the convenience of portability; a typical unit is slightly larger than a good-quality 8mm home camera—and smaller than most 16mm cameras. (With standard video equipment, you'd probably have three pieces—the camera, the portable recording unit, and a battery pack—which can be quite heavy.) In addition, the overall cost of a combined unit is lower than that of separate pieces of comparable equipment.

As this is being written, there aren't many brands or models of these units from which to choose. By the time you read the book, things will almost certainly have changed. At this writing, most single units use miniaturized cassettes and record about 20 minutes per cassette. Camcorders that use full-sized cassettes have just been released, and these use only the fastest tape speed (two hours per cassette).

FIG. A–6 Beta movie camera by Sony.

FIG. A–7 VHS camcorder by Zenith.

CAMERA MAINTENANCE

Unless you have a camcorder, there are no heads to clean in a video camera. Maintenance is a minor job, consisting primarily of protecting and storing the camera properly.

It's foolish to invest $1,000 in a quality video camera and then store it unprotected on a shelf. Spend the extra $40 or so and get a protective bag to hold your camera and its related equipment.

FIG. A–8 JVC camera bag.

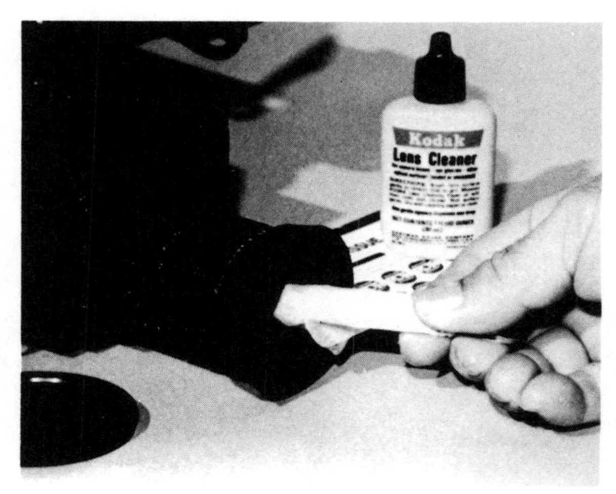

FIG. A–9 Use quality lens cleaner and paper.

Camera bags come in all sizes, shapes, strengths, and prices. You can buy cloth bags, or ones with hard casing (sometimes metal). A hard case and internal padding will provide the best protection.

The camera, in its bag, should be stored away from heat and moisture. Even when you keep the camera in a storage bag, dust and other contaminants can enter through small cracks. You can keep dirt to a minimum by cleaning the outside of the camera and peripherals regularly with a slightly damp cloth.

The most critical part of the camera to be cleaned is the lens. Use only a quality cleaner and paper, both made especially for lenses. Tissue paper, paper towels, or anything not specifically meant to clean lenses can scratch them. Cleansers designed for window glass can cause considerable damage. The same cleaning guidelines apply to other glass on the camera, such as the viewfinder.

A clean camera will not only last much longer but also take better pictures with fewer malfunctions. The circuitry inside is delicate, as are the lenses. Mistreat them and you're bound to have trouble.

CAMERA DIAGNOSTICS

Unless you have a technical background (preferably with video cameras), you should NEVER attempt to open your camera. There is no danger to you inside, but the chances are good that you could cause expensive damage to delicate components.

If there seems to be a malfunction, as usual start by checking the obvious. The reason for a blank image may be that the operator forgot to remove the lens cap or plug in the cables, notably the power cable.

If you can see through the camera but nothing appears on the screen or goes to the tape, perhaps you don't have the camera connected properly to the VCR.

Don't do anything else until you've eliminated all the obvious possibilities. Is everything connected properly? Are any batteries dead? If a power supply is used to provide the needed 12 volts DC, is it plugged in and turned on? Is it working at all?

The more complex the camera, the more likely that the operator is doing something wrong. The more attachment options the camera has, the more likely that the operator has made a mistake in hooking up something. Refer to the manual that came with the camera to check operating procedures and attachments.

You can perform many quick checks with a VOM. Batteries and power supply output can be tested by setting the VOM to the 12-volt-DC range (check your manual to confirm the setting) and touching the contacts with the probes.

Most cameras today use some version of the plumbicon, newvicon, or saticon tubes. The operation is essentially the same—light is converted into a current that can be transferred and recorded on the tape. Since these tubes are light sensitive, you have to protect them from too much light. By aiming the camera into the sun, you may cause the tube to burn out from overload.

What's more, aiming the camera at a particular object for too long can cause the image of that object to "burn in." You can usually take care of this by aiming the camera at a white card or wall or a while. If the problem still occurs, take the camera to a professional technician.

Other light-related problems, such as streaking or indistinct images are common in lower-priced equipment. If they happen under low- or bright-light conditions, the problem is likely to be that a tube cannot keep up with the changes in light intensity because it is either aged or bad.

As discussed in "Buying a Camera," seeing colored shadows indicates misalignment in the camera. If you experience this symptom, you'll probably have to send the camera back to the manufacturer for realignment. Fortunately, this problem is rare.

Cables and connectors are subject to a lot of wear. To check them, set your meter to read ohms (resistance). It helps if you already know which pins on one side go to which pins on the other, but with a bit of patience you can find out.

Since very few manuals give pin allocations, it's a good idea to run a continuity test on all multipin cables when you first get them.

Make a sketch in your owner's manual of the connector and label the pins. If something seems to have malfunctioned later, you'll know what to check. How you label the pins doesn't really matter. What counts is your determination of which pins connect to which through the wires inside the cable.

OTHER EQUIPMENT

Various other accessories are available for your VCR. Rewinders can save wear on the motors in your VCR; some models even clean the tape as it is being rewound. Control boxes make switching easier and more efficient. Editors and enhancers change and improve the recorded image.

Prices vary, again depending on the features. A simple rewinder can cost as little as $20. A rewinder that can also fast forward, clean the tape, and even bulk erase a tape can cost $150 and more. Switch boxes are often fairly high priced because of the special RF switches that are used. Likewise, the more functions an enhancer has, the more electronics there will be inside, and the more it will cost.

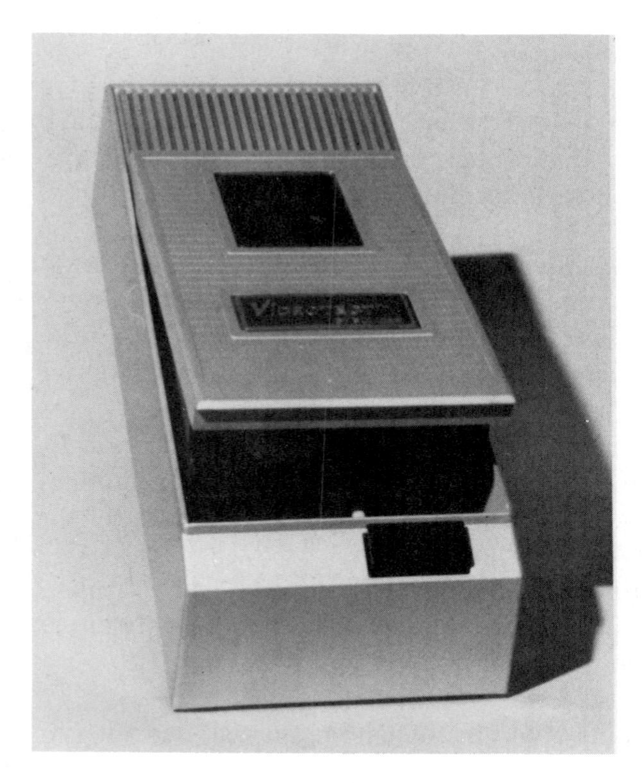

FIG. A–10 Solidex rewinder.

FIG. A–11 Panasonic switch box.

In general, a higher-priced peripheral will do more things and will probably last longer than a lower-priced one. But peripherals are not magical. If the original recording is terrible, no enhancer is going to be able to restore it to broadcast quality.

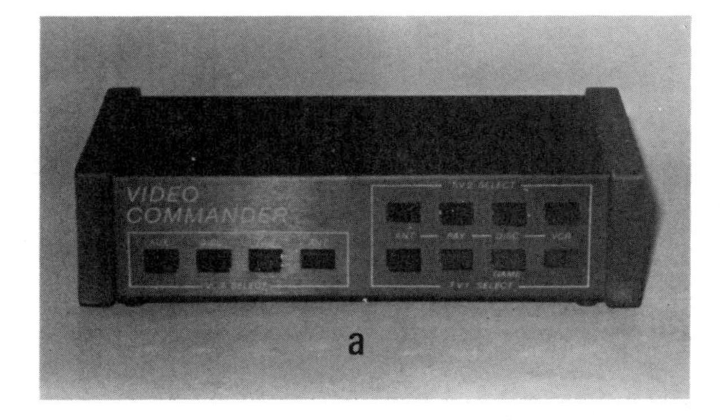

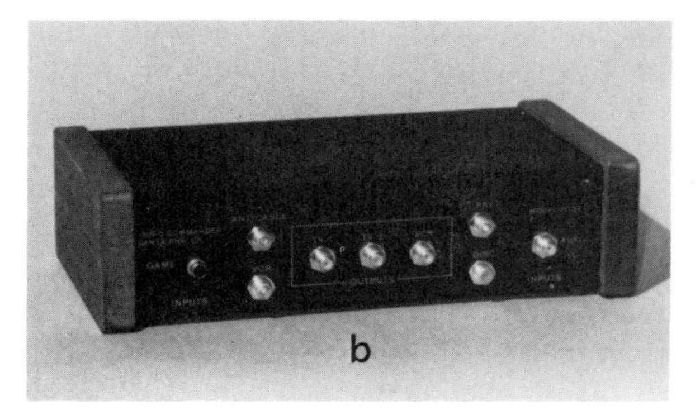

FIG. A–12 Video Commander switch box (*a*) Front (*b*) Back.

FIG. A–13 Videocraft image enchancer.

DIAGNOSTICS

As with video cameras, most peripheral equipment is made in such a way that the end user will probably not be able to repair it. Fortunately, the reliability of this equipment is generally high enough so that malfunctions are extremely rare. In most locations, you won't be able to find a shop that can do the repairs—simply because it isn't profitable for technicians to learn to do them.

As with all other equipment, diagnosis of problems is a process of elimination. If a camera or other equipment seems to be malfunctioning, just so many things can be causing the trouble. Chances are good that nothing is wrong with the equipment but the operator is doing something wrong or is expecting the equipment to do something it simply can't do. Check that all cables are connected properly and equipment plugged in. Often, the instruction manual provides a troubleshooting guide.

Next, check the cables themselves. This is done with a VOM set to read resistance. Touch the probes of the meter to the two ends of the cable, making sure that the probes touch *only* the correct ends or pins. A reading of close to zero ohms means that the wire inside is probably good. A reading of infinity means that either you've done something wrong or the wire inside has broken.

A switch that seems completely depressed (or released) may in fact not be. Try pressing it again. Check other controls as well. You may have them out of adjustment or perhaps are using them incorrectly. Put all controls to a zero setting and make your adjustments from there.

If these steps haven't solved the problem, look everything over carefully. Quite often the cause of the malfunction can be spotted visually. If not, it's probably time to consult with a professional technician.

SUMMARY

Video cameras and other equipment are tough in many respects and will operate flawlessly for years unless you mistreat them. The key is proper care and handling. When cleaning the outside of video equipment, use an only slightly damp cloth. For cleaning lenses, never use anything but lens cleaner and lens paper. In general, the more expensive the equipment, the better the quality.

Diagnosis of problems with cameras and peripheral video equipment is a matter of eliminating as many possibilities as you can. This process always begins with the most obvious things, such as operator error (e.g., maladjusted controls). Refer to the instruction manual for correct procedures.

A surprisingly common cause of trouble is the cables and connectors, which continually get jostled. Fortunately, these are about the least expensive items to repair or replace and the easiest to test (with a VOM).

Don't attempt detailed repairs—especially on cameras—unless you have experience with them. Spend the extra money to hire a professional technician: in the long run you'll be ahead.

Glossary

AC (alternating current): The incoming current from the wall outlet.

Amplifier: A device that boosts a signal.

Antenna: A device that is able to pick up a transmitted signal from the air. It is then fed through wire or cable to the VCR or television set.

Audio: The sound portion of the equipment.

Cable: The wire assembly used to carry signals, such as the RF cable that carries the signal from the VCR to the television set. Some companies, usually commercial, supply television signals to the home through the use of cable.

Capacitor: An electronic component used to store a charge or to block a DC signal. Because it allows AC signals to pass, it can filter unwanted AC signals.

Capstan: A pin, rod, or motor shaft that turns, often made of metal.

Carrier: The underlying signal that contains the information of a transmission.

Cassette: Also called a "cartridge." Both the case (or housing) and the tape used in a VCR.

CCTV: Closed-circuit television.

Common: The ground of a circuit, or the common path by which electrons return to the source (such as a battery or power supply).

Control Track: A pulse recorded on the tape that controls the video heads during playback.

CRT (cathode ray tube): The correct term for the picture tube of a television set.

DC (direct current): Current, such as that supplied by a battery, that flows in one direction only, unlike AC, which flows back and forth.

Dew Sensor: A special device in some VCRs that signals when the moisture level inside the machine is dangerously high.

Diode: A basic electronic component, often used to convert AC to DC. It allows easy flow of current in one direction but not the other.

Distortion: Impurities in the signals (noise), most often caused by interference or by flaws in the circuit.

Distribution Amplifier: A device that both boosts the signals and splits them so that they can be used by more than one television set.

Dropout: Loss of signal. A common cause is deterioration of the tape.

Dub: To transfer a recording from one tape to another.

Flagging: The apparent bending at the top and bottom of the screen.

Flutter: Distortion, usually occurring at higher frequencies.

Frame: A single picture image. Put together they give the sensation of motion.

Freon: An inert substance often used for cleaning because it does not cause a chemical reaction and it evaporates quickly with no residues.

Frequency: The speed with which AC signals shift between positive and negative. Electricity coming into the home does this 60 times per second (60 cps). A radio-frequency signal does it thousands or millions of times per second.

Ghost: An echo of the original image, caused by a reflection of the original signal.

Glitch: Interference, usually caused by relatively low frequencies, that appears as a bar moving across the picture.

Ground: See *Common.*

Helical: A spiral-shaped motion, such as the path of the videotape across the recording/playback heads.

Hooking: See *Flagging.*

IC (integrated circuit): A module containing any number of basic components, making it a sort of miniaturized circuit. Sometimes called a "chip" because it often looks like a flat block of plastic.

Monitor: The television receiver; more often, a special high-quality receiver that has no actual tuning section.

Noise: Distortion. In the audio range, distortion is an "unclean" sound. In video, it can be marked by streaks or splotches.

Ohm: Unit of electrical resistance or AC impedance.

Oxide: A sophisticated form of rust, with a chemical formula of Fe_2O_3, that allows the signal to be recorded and played back magnetically.

Peripheral: Equipment outside the main unit but connected to it, such as cameras, editors, and enhancers.

PET (polyethylene terephthalate): The technical and generic name for Mylar (DuPont). The plastic base of most recording tape.

Resistor: An electronic component that restricts the flow of current.

Resolution: The clarity of the picture.

RF (radio frequency): Much higher than the AC frequency of household current, it ranges from about 10,000 cps and up; (see *VHF* and *UHF*).

Roller: A rubber "tube" that causes the tape to move. Friction is caused by a squeezing motion of the capstan against the roller with the tape between.

Scan: To break an overall image into component parts, which are then translated electronically for storage.

Servo: A device that takes incoming power and converts it to a mechanical motion.

Signal-To-Noise (S/N) Ratio: How the wanted signal level compares to the unwanted signal level.

Skewed: Slanted from the normal motion.

Stage: A circuit within an electronic unit that performs a particular function. The term applies to amplifiers in which the signal is boosted in steps, with each step corresponding to a stage.

Sync (synchronous): A type of electronic pulse that keeps two or more devices operating in step. For example, as the camera scans a scene, a synchronous pulse ensures that the camera and monitor or VCR will scan the same spot at the same time.

Threading: Loading tape along its correct path.

Torque: The twisting force, such as of a motor.

Tracking: Keeping the recorded signal in exactly the right place for the heads to read.

Transformer: An electronic component used to convert one value of AC voltage to another.

Transistor: An electronic component that controls the flow of current similar to two diodes connected back-to-back.

UHF (ultrahigh frequency): Channels 14 to 83 on a television set or VCR tuner.

VHF (very high frequency): Channels 2 to 13 on a television set or VCR tuner.

VOM (volt-ohm-milliammeter): A tool used to test voltage and resistance.

VTR (videotape recorder): *All* machines that make video recordings. VCR describes only those that use cassettes.

Wow: Distortion in sound reproduction. Or what you say when you successfully complete a repair.

Index